AF601559

Environmental Problems and Sustainable Development

Issues and Challenges

The Editor

Dr. Roopshikha Agrawal is Assistant Professor of Botany selected by Public Service Commission of Chhattisgarh with merit. She completed her M.Sc. (Botany) from Dr. Hari Singh Gour University Sagar (M.P.) in first division with 2nd position in University merit. She has done her Ph.D. in 2002 from Dr. Hari Singh Gour University Sagar (M.P.). She has twelve year experience of teaching in Govt. Degree & P.G. Colleges of M.P & C.G. She is member of various academic societies & Ex. Co-editor of an International Journal *"Lab to Land"*. She has published several research papers in various Indian Journals. She has written a book with ISBN No.978-93-80397-21-4 on *"Fungal Foliage Diseases and Management"*.

Environmental Problems and Sustainable Development

Issues and Challenges

— Editor —

Dr. Roopshikha Agrawal

Assistant Professor, Botany
Govt. Rajeev Lochan College
Rajim (Gariaband), C.G.

2017

Scholars World

A Division of

Astral International Pvt. Ltd.

New Delhi – 110 002

ISBN: 9789387057395 (Int. Edn.)

Publisher's Note:

Every possible effort has been made to ensure that the information contained in this book is accurate at the time of going to press, and the publisher and author cannot accept responsibility for any errors or omissions, however caused. No responsibility for loss or damage occasioned to any person acting, or refraining from action, as a result of the material in this publication can be accepted by the editor, the publisher or the author. The Publisher is not associated with any product or vendor mentioned in the book. The contents of this work are intended to further general scientific research, understanding and discussion only. Readers should consult with a specialist where appropriate.

Every effort has been made to trace the owners of copyright material used in this book, if any. The author and the publisher will be grateful for any omission brought to their notice for acknowledgement in the future editions of the book.

Published by : **Scholars World**
A Division of
Astral International Pvt. Ltd.
– ISO 9001:2008 Certified Company –
4760-61/23, Ansari Road, Darya Ganj
New Delhi-110 002
Ph. 011-4354 9197, 2327 8134
E-mail: info@astralint.com
Website: www.astralint.com

Acknowledgement

The National Seminar on **Environmental Problems and Sustainable Development: Issues and Challenges** is the beginning of awareness among scientist, research scholars, intellectuals, teachers, students and society to protect Environment along with Developmental activities.

It is my great pleasure to express my gratitude to our **Chief Guest Shri Prem Prakash Pandey (**Honorable Minister Higher Education Minister, Chhattisgarh) who has given his valuable time out of his busy schedule for **Inaugural Programme** of National Seminar.

I am highly thankful to our **Special Guest Dr. S. K. Pandey** (Honorable Vice Chancellor Pt. Ravishankar Shukla University Raipur C.G.) for his valuable support and inspiration.

I extend my sincere thanks to **Dr. K. Subramaniam** (Additional Principal Conservator of forest Raipur C.G.) for his **key note** address.

I am Cordially thankful to **Resource persons of National seminar Dr. M.L. NAIK** Environmentalist and Ex. Professor Life Sciences Pt. R.S.U. Raipur (C.G.), **Dr. K. N. Sharma** (Principal Govt. College Arang C.G.), **Dr. R.V. Shukla** (Ex. Professor Botany, Ex. Principal C.M.D. College Bilaspur), **Dr. Rekha Pimpalgaonkar** (H.O.D. Botany, Govt. Nagarjun Science P.G. College Raipur C.G.), **Dr. Mayurshikha Agrawal** (Professor and In charge Principal, State Law College Bhopal M.P.), **Dr. Gunwant P. Gadekar** (H.O.D. Zoology Dhote Bandhu Science College, Gondia Maharashtra) who took pains to come **Rajim** from different states of India and given us an opportunity to share their valuable thoughts which will be provide mile stone on the way of leading Environmentalist, Professors, Students and Society.

I am Heartily Thankful to **MUKUND HAMBARDEY (Director of CG Council of Technology of Chhattisgarh)** who has very kindly consented for **Valedictory Function and Award Ceremony.**

I am Special thankful to all delegates who actively **participated** in this National Seminar.

I am thankful to **U.G.C. C.R.O. Bhopal (M.P.)** without whose actively assistance this National seminar could not be done.

I am Profoundly thankful to all who **directly or indirectly** concerned with this seminar.

R. Agrawal

Dr. Roopshikha Agrawal

Asstt. Professor (Botany)

Editorial

This is a great pleasure and honour for the **Department of Botany** of **Government Rajeev Lochan College Rajim Distt-Gariaband** for organizing a National Seminar on **Environmental Problems and Sustainable Development: Issues and Challenges** on **19-20 January 2016**. It was sponsored by **U.G.C. C.R.O. Bhopal (M.P.)**.

Earth is beautiful living planet of the Universe as the common habitat of more than 7 billion Human population and Millions of species of biodiversity. It is only possible due to its ENVIRONMENT. I feel that Nature is our home and we individual have to take up the responsibility to preserve our natural environment. Mahatma Gandhi said **"Earth provides enough to satisfy every man's needs, but not every mans greed."**

Environment and Human are interdependent on each other. It is a big challenge to maintain environmental balance along with Developmental Activities.Now a days environment is facing so many problems like pollution, global warming,climate change, depletion of ozone layer, deforestation, decreasing of ground water level of land, decrease in rain fall, segmented rain fall due to industrialization, transportation and excessive use of Natural resources, Fresh air, water and food products are not available to common man. The area of forest is gradually decreasing, some flora and fauna are about to extinction. The patients of cancer, asthma and respiratory disorder, skin diseases and disease related to nervous system are increasing day by day. The reason of all these problems to a great extent is imbalenced development. The State of Chhatiisgarh is very rich in Natural and Mineral resources. After the existence of Chhattisgarh, the state is developing very fast. There are so many industries such as iron and steel, aluminum, cement, coal and power generating plants are working. But the environment is also being polluted. Hence in order to save environment from these problems the policies, Laws, legistation, technology, Government and the common man should come forward. This should be an integrated efforts.

National Seminar had Provided Platform to Educational Administrators, Vice -Chancellors, College Principal, Deans, Readers, Head of Departments, Professors,

Assistant-Professors, Scientists, Environmentalist, Researchers, Young Scientists and Students to disseminate knowledge related to ***"Environmental Problems and Sustainable Development: Issues and Challenges"***. The general topics covered in the National Seminar are as under:

- ✰ Environmental Pollution, Health and Management.
- ✰ Documentation and Conservation of Flora and Fauna, Development of Green Fuel
- ✰ Usage of Bio Remediation for Cleaning Environment
- ✰ Solid Waste Management
- ✰ Environmental Policies, Laws and Legislation

Today, Environment education very important from primary level to higher level of education, So I am thankful to **U.G.C.**that Environment subject is added in syllabus as a compulsory subject in all colleges, because Environment subject is related with all subjects.

Dr. Roopshikha Agrawal
Asstt. Professor (Botany)

बलरामजी दास टंडन
राज्यपाल छत्तीसगढ़

राजभवन
रायपुर - 492001
छत्तीसगढ़
भारत
फोन : +91-771-2331100
+91-771-2331105
फैक्स : +91-771-2331108

क्र. 05/पीआरओ/रास/16
रायपुर, दिनांक 08 जनवरी 2016

संदेश

प्रसन्नता का विषय है कि शासकीय राजीव लोचन महाविद्यालय, राजिम द्वारा **'पर्यावरणीय समस्याएं एवं सतत् विकास : मुद्दे एवं चुनौतियां'** विषय पर दो दिवसीय राष्ट्रीय सेमिनार के आयोजन के साथ ही संक्षेपिका का भी प्रकाशन किया जा रहा है।

आज विकास की अंधाधुंध दौड़ में हम पर्यावरण को नजरअंदाज करते जा रहे हैं। यदि समय रहते पर्यावरण का सुनियोजित एवं सुसंगत प्रबंधन नहीं किया गया तो इसके दुष्परिणाम सम्पूर्ण मानव जाति को भुगतने होंगे।

मुझे आशा है कि महाविद्यालय द्वारा आयोजित सेमीनार एवं प्रकाशन से पर्यावरण प्रबंधन के क्षेत्र में आम जनता में जागरूकता आएगी और पर्यावरण संरक्षण के क्षेत्र में कार्य करने की प्रेरणा मिलेगी।

हार्दिक शुभकामनाओं सहित।

(बलरामजी दास टंडन)

डॉ. रमन सिंह
मुख्यमंत्री

Dr. RAMAN SINGH
CHIEF MINISTER

DO. No. 156 / VIP / 2015
DATE 06/01/2015
महानदी भवन, मंत्रालय
नया रायपुर, छत्तीसगढ़ - 492002
Mahanadi Bhawan, Mantralaya
Naya Raipur, Chhattisgarh
Ph. : (O) - 0771-2221000-01
Fax : (O)- 0771-2221306
Ph. : (R) - 0771-2331000-01
Ph. : (R) - 0771-2443399
Ph. : (R) - 0771-2331000

संदेश

मुझे यह जानकर हार्दिक प्रसन्नता हुई कि शासकीय राजीव लोचन महाविद्यालय, राजिम, जिला गरियाबंद द्वारा 'पर्यावरणीय समस्याएं एवं सतत विकास : मुद्दे एवं चुनौतियां' विषय पर राष्ट्रीय संगोष्ठी का आयोजन किया जा रहा है। नियमित अध्ययन–अध्यापन के अतिरिक्त विषय विशेषज्ञों के साथ विचार–विमर्श का अवसर मिलना छात्र–छात्राओं के साथ ही संस्था से जुड़े सदस्यों के लिए भी महत्वपूर्ण होता है। ऐसे आयोजनों से काफी उपयोगी जानकारियां प्राप्त होती हैं।

आयोजन तथा प्रकाशन की सफलता के लिए मेरी शुभकामनाएं।

(डॉ. रमन सिंह)

प्रेम प्रकाश पाण्डेय
मंत्री,
राजस्व एवं आपदा प्रबंधन, पुनर्वास, उच्च शिक्षा,
तकनीकी शिक्षा एवं जनशक्ति नियोजन,
विज्ञान एवं प्रौद्योगिकी विभाग, छत्तीसगढ़ शासन.

(मंत्रा.) : 0771-2510321, 2221321
(निवा.) : 0771-2446440
: बी-1, शंकर नगर, रायपुर (छ.ग.)

निवास : सेक्टर-9, सड़क नं.-11,
बंगला नंबर-1, भिलाई (छ.ग.)
: 0788-2242590

क्रमांक 10 मंत्री/रा.आ.प्र उ शि.त.शि.ज नि.वि.प्रौ./2014-15 रायपुर, दिनांक 04/01/2016

Message

I am glad to know that the Department of Botany, Govt. Rajeev Lochan College, Rajim, Distt Gariyaband (C.G.) is going to organize a National Research Seminar on "Environmental Problems and Sustainable Development : Issues and Challenges", sponsored by U.G.C., C.R.O. Bhopal (M.P.)

To face and overcome the development related challenges of 21st century, our society should be more aware of the environmental issues. It is very important to inculcate, within each one, the feeling of devotion towards perseverance of the environment and the natural resources.

I hope that authentic research papers presented in the National Seminar organized on such a topic will not only benefit the education world, but also the public in general.

My best wishes for success of the seminar and publish of proceedings.

(Prem Prakash Pandey)

To,

Principal,
Govt. Rajeev Lochan College,
Rajim, Distt Gariyaband (C.G.)

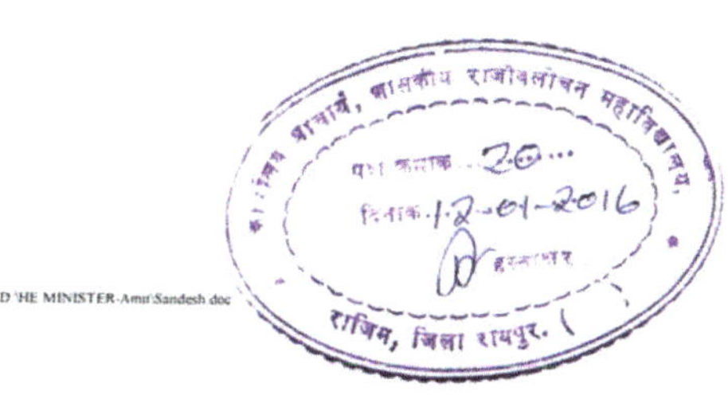

D:\HE MINISTER-Amit\Sandesh.doc

राजेश मूणत
मंत्री
लोक निर्माण,
आवास एवं पर्यावरण, परिवहन
छत्तीसगढ़ शासन

फोन : (कार्या.) 2221316, 2510316
(निवास) 2331065–66

फैक्स : (कार्या.) 0771–2221316
(निवास) 0771–2331065

क्रमांक : Q/V.Z.P./2015-16

रायपुर, दिनांक : 31/12/2015

शुभकामना संदेश

मुझे यह जानकर प्रसन्नता हुई कि शासकीय राजीव लोचन महाविद्यालय, राजिम, जिला–गरियाबंद द्वारा ''**पर्यावरणीय समस्याएं एवं सतत विकास : मुद्दे एवं चुनौतियां**'' विषय पर राष्ट्रीय सेमीनार का आयोजन किया जा रहा है। सेमीनार पर्यावरण समस्याओं को दूर करने में सहायक सिद्ध होगा। साथ ही इस अवसर पर प्रकाशित स्मारिका अपने उद्देश्य मे सार्थक सिद्ध होगी।

हार्दिक शुभकामनाओं के सहित........

(राजेश मूणत)

डॉ. शिवकुमार पाण्डेय
कुलपति

Dr. S.K.Pandey
Vice-chancellor

पं. रविशंकर शुक्ल विश्वविद्यालय, रायपुर (छ.ग.), भारत
Pt. Ravishankar Shukla University, Raipur (C.G.), INDIA
Office : +91-771-2262857, Fax : +91-771-2263439
E-mail : vc_raipur@prsu.org.in
Website : www.prsu.ac.in

स्वर्ण जयंती वर्ष Golden Jubilee Year 2013-14

रायपुर, दिनांक 05 जनवरी, 2016

शुभकामना

हार्दिक प्रसन्नता का विषय है कि शासकीय राजीव लोचन महाविद्यालय राजिम में यू.जी.सी., क्षेत्रीय कार्यालय भोपाल के सौजन्य से **"पर्यावरणीय समस्याएँ एवं सतत विकास : मुद्दे एवं चुनौतियाँ"** विषय पर दिनांक 19 एवं 20 जनवरी 2016 को दो–दिवसीय राष्ट्रीय सेमीनार का आयोजन किया जा रहा है।

वर्तमान में भारत ही नहीं अपितु सारा विश्व अनेक पर्यावरणीय समस्याओं का सामना कर रहा है। इन समस्याओं का समाधान पर्यावरण प्रबंधन से ही संभव है, जिसके लिए हर जगह व हर एक के द्वारा संगठित प्रयास आवश्यक है।

आशा है कि इस सेमीनार से सभी वर्गों के मध्य संवाद के अवसर उपलब्ध होंगे तथा विषय विशेषज्ञों द्वारा प्रस्तुत तथ्यों से जनसामान्य लाभान्वित होंगे। मैं इस राष्ट्रीय सेमीनार के सफल आयोजन की कामना करता हूँ।

शुभकामना सहित।

(डॉ.एस.के.पाण्डेय)

संतोष उपाध्याय
विधायक
विधानसभा क्षेत्र क्र.-54
राजिम

तहसील कार्यालय के पास,
फिंगेश्वर रोड, राजिम
जिला - गरियाबंद (छ.ग.)
मोबा. : 099773-13101

क्रमांक : 284 दिनांक : ....

संदेश

अत्यंत हर्ष की बात है कि शास. राजीव लोचन महाविद्यालय राजिम द्वारा "**पर्यावरण समस्या एवं सतत् विकास : मुद्दे व चुनौतिया**" विषय पर दो दिवसीय राष्ट्रिय संगोष्ठी का आयोजन किया जा रहा है।

वर्तमान में वायु, जल, भूमि प्रदुषण, प्राकृतिक संसाधनों का अंधाधुंध दोहन तथा बढ़ता तापक्रम आदि ऐसे खतरे है जिससे सम्पूर्ण विश्व चिंतित है, यह चिंता का विषय पर्यावरण से जुड़ा हुआ है तथा इसका परिणाम बहुत ही विनाशकारी हो सकता है। अतः इसके लिए प्रत्येक जन को जागरूक एवं शिक्षित होना आवश्यक है।

मेरा मानना है कि इस प्रकार के सेमिनार सभी वर्गो के मध्य संवाद का बेहतरीन माध्यम है, मै इस राष्ट्रिय सेमिनार के सफल आयोजन की कामना करता हूँ।

(संतोष उपाध्याय)

जितेंद्र सोनकर
अध्यक्ष जनभागीदारी समिति
शासकीय राजीव लोचन महाविद्यालय राजिम
जिला- गरियाबंद (छ. ग.)

संदेश

अत्यंत हर्ष का विषय है कि शासकीय राजीव लोचन महाविद्यालय, राजिम का वनस्पति विभाग विश्वविद्यालय अनुदान आयोग सी.आर.ओ भोपाल के सहयोग से 'पर्यावरणीय समस्याएं एवं सतत् विकास मुद्दे एवं चुनौतियां' विषय पर दो दिवसीय राष्ट्रीय सेमीनार का आयोजन कर रहा है।

राजिम क्षेत्र के लिए इस प्रकार के आयेजन एवं प्रकाशन हेतु आयोजकगण बधाई के पात्र है।

राष्ट्रीय सेमीनार की सफलता के लिए मेरी हार्दिक शुभकामनाएं

(जितेंद्र सोनकर)

नगर पंचायत, राजिम

जिला – गरियाबंद छ.ग.

पवन सोनकर
अध्यक्ष, नगर पंचायत, राजिम
जिला – गरियाबंद (छ.ग.)

संदेश

यह अत्यंत प्रसन्नता का विषय है, कि यू.जी.सी. सी.आर.ओ. भोपाल द्वारा प्रायोजित एवं वनस्पति विभाग शासकीय राजीव लोचन महाविद्यालय, राजिम द्वारा **"पर्यावरणीय समस्या एवं सत्त विकास : मुद्दे एवं चुनौतियॉ"** विषय पर दो दिवसीय राष्ट्रीय संगोष्ठी का आयोजन किया जा रहा है ।

वर्तमान परिवेश में पर्यावरण संबंधी जागरूकता एवं संविकास एक बहुत बड़ी आवश्यकता बन गई है और इस मुद्दे पर पूरी दुनिया में निरन्तर विचार - विमर्श और नये कदम उठाया जाना आवश्यक है।

मै आशा करता हूँ कि इस संगोष्ठी के माध्यम से पर्यावरण समस्या एवं सतत् विकास पर गहन विचार विमर्श किया जाएगा एवं सार्थक परिणाम सामने आयेंगे जिससे मानव समाज को एक नई दिशा प्राप्त होगी।

यह संगोष्ठी सफल हो इसके लिए मेरी हार्दिक शुभकामनाऍं।

पवन सोनकर

Preface

Environment may be defined in the simple terms as the sum total of all external conditions and influences that affect the living organisms. It includes lower part of atmosphere, entire hydrosphere, and lithosphere, where existence of living organism has been found.

Our environment is constantly changing. There is no denying that however, as our environment changes, so does the need to become increasingly aware of the problems that surround it with massive influx of natural disasters, warming and cooling periods, different types of weather patterns and much more, people need to be aware of what type of environmental problems our planet is facing. Every day the population issue is going worst and worst it more like slow poison which is hardly visible through our naked eyes. Pollution has only one role is to destroy our ecosystem and environment slowly with time.

There are numerous things that everyone can do at home help mitigate this environmental damage. Changes can add up to a major shift. So we should follow some small steps as, use compact fluorescent light bulbs. It is true that these bulbs are more expensive but they can save energy and in the long term our electricity bill would be reduced. driving is one of the biggest causes of pollution, So we should walk or use bike if the journey is short one. We should save the rain water. This water can be used for different purposes, all of these steps to protect our environment.

If we want to protect our environment from pollution, then we will have to aware all people.

Therefore, I am thankful to organizer to organize this National Seminar.

Dr. Priti Tiwari

Principal

Contents

Impact of Toxicant of Fertilizer Industry Waste Effluent at Bhopal (M.P.)

Vibha Choubey[1] and Parmita Dubey[2]

[1]*Department of Zoology, Govt. B.P.A.C.S. College Arang, Raipur, C.G.*

[2]*Govt. J.Y. Chhattisgarh College, Raipur, C.G.*

ABSTRACT

The in-plant pollution control measures are very significant particularly in case of Fertilizer industry. The relates to elimination in volume and strength of effluent by incorporation of suitable pollution control measures in the plant itself. The studies on Bio-Physico-Chemical analysis of the Bhopal city were carried out from December 2010 – 2011. It has been concluded that Bhopal Lake is highly eutrophic and biologically 'dead' in term of its un ability to provide the aesthetic pleasures of swimming, boating, fishing and the effluent of fertilizer industries due to luxuriant growth of micro and macro flora and fauna.

Keywords: *Water quality, Eutrophic, Physico-chemical, Organic pollution, Fertilizers.*

Introduction

Bhopal is the capital town of the M.P. Its geographical area is 284.9 km^2 (Census report, 1991) and its altitude above sea is 505 m. The city lies between Latitude 23° 34 inches N and longitude 77° 10 inches to 77° 10 inches to E.

The Bhopal Lake is facing the problem of organic pollution due to the major discharge of effluent of fertilizer industries, raw sewage, and detergents etc. Lake emanates foul smell in summer season due to effluent accumulation there by causing health hazards to the surrounding human population. Hence the present work was undertaken to evaluate the status of organic pollution and water quality by Bio-Physico-Chemical analysis. Some notable work on this aspect has been done by Dhamija and Jain (1994), Kumar (1995), Rao *et al.* (1996), Khare (1998, 1999) and Kumar and Singh (1998), Tripathi (2007).

Materials and Methods

Water sample were collected twice in a month of years 2010-2011, They were analyzed for Physico-Chemical (Adoni, 1985, Trivedi and Goal, 1986) and Biological parameters (Michael, 1973 and Adoni, 1985). Plankton samples were collected by standard methods from predetermined sampling sites and preserved in 2-5 per cent formaline and few drops of glycerin. Counting and identification of plankton were done as per/Apha (1985), Adoni (1985) and Michael (1973).

Results and Discussion

The result of plankton population is shown in (Tables 1.1 and 1.2). The temperature was higher in month of April 2011, lower in December 2010 and medium in February 2011. The range being between 18.-35 C. it has an indirect effecting. The toxicity, intensifying deoxygenating and finally increasing the biomagnifications that is why, the dissolved oxygen depletion and plankton community intensity their span in month of April 2011. Eutrophic waters are characterized by blooms of Cynobacteria (Kumar, 1995) this is true in case of Bhopal lake. The nature and health of the aquatic communities is an expression of the quality of water (APHA, 1985). Water pollution manifest through changes such as Physical, Chemical and Biological.

Lake water was alkaline through out the study period (7.1-11.1). The alkalinity was lower in the month of December (2010), higher in April (2011). The minimum and the maximum values of the total dissolved solids were 460 to 610 in Table 1.3.

Table 1.1: Phytoplankton Population Observed in Bhopal Lake Water during (2010-2011)

Class Chlorophyceae	*Class Cyanophyceae*
1. *Eudorina* sp.	1. *Microcystis aeruginosa* kunt=
2. *Chlorella*	2. *Micro cystiscyanea*
3. *P. duplex* megan	3. *Anabaena circinalis*
4. *P. simplex* meger	4. *Rivularia gigantean* Schmilde
5. *Spirogyra* sp.	5. *Nostac* sp.
6. *Mougeotia transeque* Collins	
7. *Mougeotia gelatinosa* Wittrock	*Class Euglenophyceae*
8. *Closterium* sp.	1. *Phacus platalac* Dreg.
9. *Casmarium* sp.	

Table 1.2: List of Zooplankton Population Observed in Bhopal Lake, Bhopal (M.P.)

Phylum: Protozoa	*Phylum: Arthopoda*
1. *Amoeba*	1. *Basmanyia* sp.
2. *Paramecium aurelia*	2. *Daphnia similes*
3. *Vorticella campanula*	3. *Moina* sp.
4. *Euglena* sp.	4. *Monostyla* sp.
Phylum: Rotifera	5. *Nauplius* larvae
1. *Asplanchna* sp.	6. *Cyclops viridis*
2. *Brachionus falcatum*	
3. *Keratella tropica*	
4. *Lecane* sp.	

Table 1.3: Physico-Chemical Characteristics of Water of Bhopal (M.P.) from December 2010-to April 2011

Parameters	*Dec. (2010)*	*Jan. (2011)*	*Feb. (2011)*	*Mar. (2011)*	*Apr. (2011)*
Temperature 0°C	18.5	17.6	21.2	25.9	34.1
Transparency in cm	22.4	17.2	19.6	24.0	27
pH	7.2	7.8	8.8	8.6	11.0
Total dissolved solid (ppm)	580	568	490	460	610
DO (ppm)	12.8	14.6	11.8	8.9	7
Free CO_2 in (ppm)	3.00	6.2	7.05	5.6	4.3
Carbonate (ppm)	78.7	110.4	60.4	88.2	100
Ca in ppm	29.0	33.3	34.4	37.1	40.8
Mg in ppm	12.4	12.0	12.4	14.7	21.6
Phosphate in ppm	0.11	0.10	0.33	0.20	0.22
Nitrate inppm	1.37	1.16	1.01	1.10	0.84
Chloride in ppm	44	42.5	58	65	66.7
Potassium in ppm	91.8	98	100	93.2	102.6

The algal population has a direct relationship with the total solids (Rao et. al,1996; Khare, 1999), which was also reflected in the Bhopal Lake investigation. The all date of Tables 1.1–1.3 clearly indicated that the water is facially contaminated by sewage and fertilizer high alkaline, enriched with nutrients favors the microbial growth and render it unsuitable for portable purpose. The sudden depletion of oxygen is the main cause for heavy mortality of organisms (April). Overstocking of organisms may also be described to the large scale destruction during anoxic condition of Lake water in such a situation even a small fluctuation in dissolved oxygen will produce adverse effect. Thus, It has been concluded that the Bhopal Lake is highly eutrophic and biologically dead/because the total production of biomass is many times greater in eutrophic lake than oligotrophic tank.

References

Adoni; A.D. 1985. Workbook on Limnology, Pratibha Publication, Sagar (M.P.); 1-212.

APHA 1985 Standard methods for the examination water and wastewater. 16th Edition Washington.

Dhamijia, S.K. and Jain, Y. 1994. Variation in the Physico-Chemical Characteristics of a lentic water body of Jabalpur M.P. *Jor. Env. Pollu.* 1; 125 – 128.

Kumar, A. 1995. Periodicity and abundance of plankton in relation to Physico Chemical characteristics of a tropical wetland of South Bihar. *Eco. Env. 8 Cons* 1; 47-51

Khare, P.K. 1990. Limnological study during rainy season of Jagat Sager Pond, Chhatarpur, M.P. *Eco. Env. 2 Cons.* 4; 279-280.

Kumar, A and Singh, R.N.P. 1998. Biodiversity and pollution status of Masanjore reservoir (South Bihar) in relation abiotic factors. *J.Eco. Env. 2Cons* 4; 139-144

Khare, P.K. 1999. Phytoplankton as indicator of water quality and pollution status of Jagot Sagar Pond, Chhattarpur (M.P.) *Geobios New reports.* 18; 107-110

Michael.R.d. 1973. Guide to the study of fresh water organism Dptt. of Biological Scinces, Madurai University, Madurai.

Rao, A.M.M. Rao. V.N. and Mahmood, S.K. 1996. Assessment of Water quality and pollution of Nursing Pond. *Jou. Env. 2 Cons.* 2; 45—49

Trivedi, R.K. and Goel, P.K. 1986. Chemical and Biological methods for water pollution studies environment publication Karad, India.

Tripathi, A 2007. Analysis of fertilize industry at Bhopal.

2

Water Quality of Tandula River in Balod Area, District Balod, Chhattisgarh

E.P. Chelak[1] *and K. Sharma*[2]

[1]*Lt. R.V.B.S. Govt. College Saraipali, Mahasamund, Chhattisgarh*
[2]*Department of Botany, Govt. Arts and Commerce Girls College, Raipur*

ABSTRACT

The water samples collected from Tendula river in Balod Area District Balod Chhattisgarh., India, were analyzed for physical properties like color, temperature, turbidity and odor, chemical properties like pH, alkalinity, total hardness, calcium hardness, magnesium hardness, total solids {Total dissolved solids (TDS)}, sulphates, nitrates were analyzed. Microbiological characteristics like detection of coliforms, quantitative analysis and most probable number (MPN) of coliforms was also performed. Incidences of Escherichia coli 0157 (Thermo tolerant strain) and Salmonella species were analyzed. All stated properties were analyzed for both upstream and downstream sampling points to determine the effect of residential and industrial discharges on the quality of river water. The addition of discharges has shown many fold increase in all the analyzed physiochemical parameters. The MPN/100 mL for upstream sample was 900 and downstream sample showed rises up to 1600. The heterotrophic plate count (HPC) also increased from 1.30x104/100 mL to 1.53x104/100 mL. Incidences of E.coli 0157 (Thermo tolerant strain) and Salmonella species were both found even before the addition of discharges.

Keywords: *MPN, E. coli 0157, TDS and TSS, River water quality.*

Introduction

Water is the most basic and vital resource of our planet. According to the UN (United Nations) reports, 1978 consumable water levels are up to 2.7 per cent of the total water content. 1 per cent of the ground water levels are threatened either directly or indirectly by pollution[1]. River water acts as receiving end of industrial wastes, residential area discharges which increase the pollution stress on these surface water bodies[2]. Non pathogenic faecal organisms are best indicators of faecal pollution. However, in all cases faecal coliform counts and *Escherichia coli* is used as the major tool in the assessment of the health risks borne by pathogens in water[3]. Many researchers constantly undertake work on assessment of physiochemical and microbiological quality of water bodies in different parts of the country[4–7]. In Chattisgarh state many small and big rivers co-originate. Tandula river is main tributary of river Shivnath. The present study focuses on assessment of impact of the discharges on the physicochemical and microbiological quality of Tandula river water in Balod Area Dist. Balod Chhattisgarh.

Experimental

River water samples were collected at two different sites hence forth cited as upstream and downstream. Tandula River enters Balod; the spot chosen for upstream sampling was at Tandula Bridge, near Ganga Maiya Temple, Jhlmala Balod. The city Channel which is one of the major sources of residential and industrial area discharges mixes into the river at Muktidham Balod, is referred as downstream. Collection of water samples for physio chemical analysis Five liters of grab samples were collected. The samples were collected in plastic containers. The collected samples were analyzed for physiochemical analysis. Collection of water samples for microbiological analysis The water samples were collected in clean, sterilized narrow mouthed glass bottle of 250 mL capacity.The collected samples were analyzed for microbiological parameters. Methodology APHA[8] (American Public health Association) was used for analysis. The method used for studying physical properties like color was by visual observation and odor by smelling, temperature by thermometer (0°-1000° C), turbidity by turbidometer, chemical properties like pH by pH meter, DO, BOD, COD, alkalinity, total hardness, calcium hardness, magnesium hardness by titrometric methods, total solids (TDS) by gravimetric method, sulphates, nitrates were analyzed by colorimetric method. Microbiological properties like detection of coliforms, quantitative analysis (MPN) of coliforms, incidences of *E.coli* 0157(Thermo tolerant strain) and *Salmonella* species were analyzed for both upstream and downstream sampling. All the chemicals used were of analytical grade.

Results and Discussion

The study revealed that, addition of discharges into the river water has enhanced the pollution load of all the parameter taken under study like pH, Temperature, Turbidity, TDS, TSS, alkalinity, hardness, BOD (Biological Oxygen Demand), COD (Chemical Oxygen Demand), nitrates, nitrites and sulphates. The results of analysis are also represented in Table 2.1 and graphical manner.

Observations of the stated parameters are given in Table 2.1.

pH, Temperature, Turbidity, TDS

The upstream and downstream water samples were found to be colorless and odorless with 23.8 (°C) and 25.2 (°C) temperature respectively. The permissible limit

Table 2.1: Physio-chemical Analysis of Water Samples from Tandula River in Balod Area Dist. Balod

Sl.No.	*Parameters*	*Upstream*	*Downstream*	*BIS[9] (Tolerance Limit) Desirable*	*BIS[9] (Tolerance limit) Maximum*
1.	pH	8	9	6.5	8.5
2.	Temperature	24.8	26.5	–	–
3.	Turbidity (NTU)	6	12	5	10
4.	TDS	420	680	500	2000
5.	DO	6.89	5.02	4	–
6.	BOD	8	15	3	–
7.	COD	14	20	–	–
8.	Ca- hardness	60	85	75	200
9.	Mg -hardness	180	260	200	600
10.	Sulphates	4O	10	200	400
11.	Nitrates	20	60	50	–
12.	Nitrites	4	15	–	–
13.	Alkalinity, mg/L	180	290	200	600

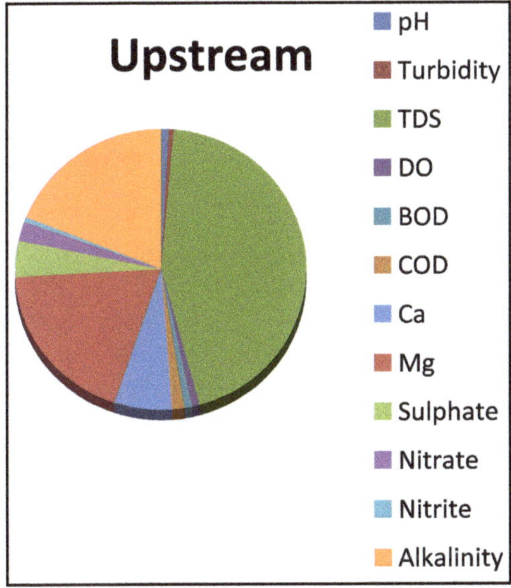

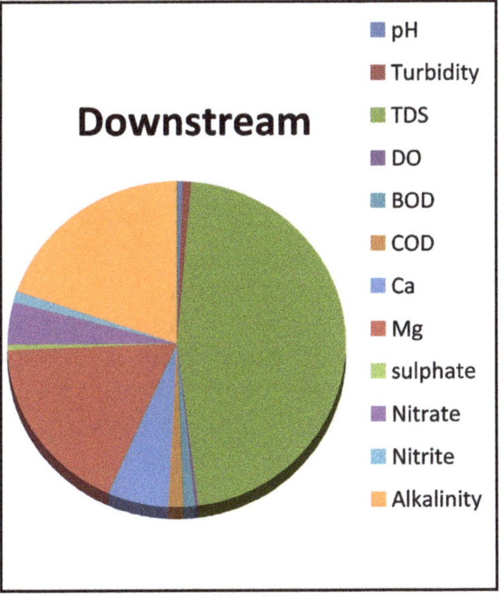

for pH is 6.5 to 8.5 for use of the river water for drinking purpose. The addition of the discharges resulted in increase of the pH beyond the BIS[9] (Bureau of Indian Standards) tolerance limit. The values of TDS were found to be within limits even after the effluent introduction into the river.

DO, BOD and COD

The BOD values of both the samples were observed to be quite high than the BIS tolerance limit indicating high organic matter content in river water.

Hardness, Sulphates, Nitrates and Nitrites

The introduction of the effluent caused the nitrate of river water to exceed the tolerance limit. According to the BIS 10500: 1991 the desirable limit of calcium should be 75 mg/mL and the permissible limit is 400 mg/mL. The calcium content of river water was observed to exceed the desirable limit but was within permissible limit.

Alkalinity

The river water shows increase in alkalinity beyond the limit. This is attributed to the routine washing and bathing activities conducted at the banks of the river.

Microbiological Analysis

Standards of acceptable total coliforms in inland surface water, according to Indian Standards 2296[10] (Indian standards of drinking water-specification BIS 10500: 1991).

Classification	*Uses Acceptable Standard*
A. Drinking water without conventional treatment, but after disinfecting	50 or less
B. Outdoor bathing	500
C. Drinking water source with conventional treatment followed by disinfecting	5000
D. Propagation of wildlife and fisheries	No standard
E. Irrigation, industrial cooling and controlled waste disposal	No standard

Microbiological studies revealed that the effluents induced microbial growth in the river water by increasing the MPN from 900 to 1600/100 mL.The HPC of upstream river water sample was 1.30x104/100 mL.The downstream sample revealed the increased value of 1.53x104/100 mL. Incidences of *Salmonella* spp. and 0157 *E.coli* were found in both upstream and downstream samples. According to the BIS standards the water is unsuitable for bathing. Raw river water cannot be used for drinking purpose.

Conclusion

The above studies reveal that there was slight increase in physiochemical parameters.The BOD, calcium content and alkalinity exceeded the BIS limits. Significant impact was found on microbiological characteristics. The MPN of coliforms nearly doubled. The quality of river water deteriorated and cannot be used for bathing purpose.The river water is unsuitable for drinking without any disinfection.

References

1. Davis M and Cornwell DA,Introduction to Environmental Engineering. Mc.Graw Hill.PWS,Publishers, New York, 1991, 93.
2. Ajayi S O and Ossian O, *Environ Pollut.* (*Series B*), 1981, 2, 87-95.
3. Byamukama D, Kansiime F, Mach R L and Farnleitner A H, *Appl Environ Microbiol.*,2000, 66, 864-868.
4. Nath D, *J Inland Fish Soc India*, 2001, 33(2), 37-41.
5. Vijender Singh, *Res J Chem Environ.*, 2006, 10(3), 62-66.
6. Mishra A and Bhatt V, *E Journal of Chemistry*, 2008, 5(3), 487-492
7. Ramteke P N, Battacharjee J W, Pathak S P and Karla N, *J Appl Bacteriol.*, 1992,72, 352.
8. Standard methods for the examination of water and wastewater (20th Ed.). Washington DC, USA. APHA (American Public Health Association), AWWA (American Water Work Association) and WEF (Water Environment Federation), Washington DC, 1999.
9. Tolerance limits for inland surface waters subject to pollution. Bureau of Indian Standard, New Delhi, (BIS: 2296-1982.), 1982.
10. Indian standards of drinking water-specification, BIS 10500: 1991.

3

Degradation of Some Industrial Dyes by Different Basidiomycetous Fungal Species

Chandrabas Sahu and Bharati Verma

Department of Biotechnology,
Seth Phoolchand Agrawal Smrity Mahavidyalaya, Nawapara, Raipur

ABSTRACT

Industrial effluents are one of the major causes of pollution especially in causing water pollution. The waste waters discharged from textile industries are the most pollutant wastes both as volume and waste composition which causes the biggest ecological problems. Out of many contaminants present in wastewater colors are considered the most undesirable mainly caused by dyes. Dyes are synthetic aromatic water-soluble dispersible organic colorants, having potential application in various industries. Over 10000 dyes are commercially available and 10-15 per cent of all dyestuff produced, are released directly to the environment and become part of the wastewater (i.e. 7×5 tons dyestuff produced annually, worldwide).

Recent fundamental works have shown the existence of many microorganisms capable of decolourizing wide range of dyes. In my present work i had done my work with seven species of basidiomycetes which were Pleurotus florida, Pleurotus sajor-caju, Polyporus sp. 1, Jelly sp., Schizophylum commune, Poly PMB., Maitake. These species were used for the degradation of malachite green and congo red. After incubation period, only one species Pleurotus sajor-caju showed degradation of congo red but 6 species are not showed degradation of congo red. Whereas 5 of the species showed degradation of congo red which were Pleurotus sajor-caju 30.5 per cent under in 10 days Pleurotus sajor-caju 77.5 per cent repetitively. Whereas 5 of the species showed degradation of malachite green which were Pleurotus florida, Pleurotus sajor-caju, Polyporus sp. 1, Jelly sp., Schizophylum commune The amount of dye degraded in 5 days incubation were Pleurotus florida 12.99 per cent, Pleurotus sajor-

caju 14.03 per cent, Polyporus sp. 1 14.16 per cent, Jelly sp.45 per cent, Schizophylum commune13.80 per cent under in 10 days Pleurotus florida 53 per cent, Pleurotus sajor-caju 47.18 per cent, Polyporus sp.1 48.16 per cent, Jelly sp. 88 per cent, Schizophylum commune 47.93 per cent repetitively.

Kyewords: *Industrial dyes, Degradation, Fungal spp.*

Introduction

Dyes are synthetic aromatic water-soluble dispersible organic colorants, having potential application in various industries. The dyestuff usage has been increased day by day because of tremendous increase of industrialization and man's urge for color.The effluents of the industries are highly colored and the disposal of these wastes into receiving water causes damage to the environment.

Materials and Methods

Culture Collection

Seven cultures collected from modern biotech lab are:

Pleurotus florida, Pleurotus sajor-caju, Polyporus sp. 1, *Jelly* sp., *Schizophylum commune,* Poly PMB., Maitake

Reviving of Pure Culture

By two methods we can revive:

A. Potato Dextrose Agar

B. Slant preparation

Potato Dextrose Agar: Media Composition and Preparation

Peeled potato tubers 200 g, Dextrose 20 g, Agar 20 g, Distilled water 1000 ml, pH 5.6.

Procedure

500 ml of water in one liter beaker was taken.200 g of washed, peeled and sliced potatoes were added to the beaker. Potatoes were gently boiled for 30 minutes or the time till they were easily penetrated by a glass rod. Filtered through cheese cloth, squeezing out all liquid.20 g dextrose was added to the potato extract.500 ml of water was taken in another beaker and was heated. 20 g agar was added, bit by bit to the hot water to dissolve it. Agar was mixed with the potato extract. Volume was making up to 1000 ml by the addition of distilled water. Media was added to conical flask and cotton plug was applied. Autoclaved at 121 C, 15 lbs pressure. Flask was allowed to cool until the flask can be held by hand. Medium was poured into petri dishes quickly under aseptic condition.

Slant Preparation

Procedure

The tubes having cotton plug were sterilized. The media was prepared and autoclaved. Test tubes and media were taken in laminar air flow and the media was poured in the tubes. Agar tubes were kept in the slanting position till the medium filled is set to form a gel or it solidifies.

Screening for Dye Degradation

Primary Screening

Screening test for dye degradation was carried out by the method of Rania M.A. Abedin, 2008. A disc(6 mm) of fungal mycelium in PDA was inoculated into the centre of petri-dish (90 mm) with the previously mentioned culture medium with agar.The plates were incubated at $30^{\circ}C$ for 3 days.The decolourization and growth were observed. Plates containing the dye but not inoculated served as control.

Dyes - 0.001 per cent 1. Malachite green, 2. Congo red, Distilled 1000 ml, pH 5.6.

Procedures

500 ml of water in one litre beaker was taken. 200 g of washed, peeled and sliced potatoes were added to the beaker. Potatoes were gently boiled for 30 minutes or the time till they were easily penetrated by a glass rod. Filtered through cheese cloth, squeezing out all liquid. 20 g dextrose was added to the potato extract. 500 ml of water was taken in another beaker and was heated. 20 g agar was added, bit by bit to the hot water to dissolve it. Agar was mixed with the potato extract. Volume was making up to 1000 ml by the addition of distilled water. Media was divided into two 500 ml conical flask equally and dyes were mixed in it separately and cotton plug was applied. Autoclaved at 121 C, 15 lbs pressure. Flask was allowed to cool until the flask can be held by hand. Medium was poured into petri dishes quickly under aseptic condition. Observation was taken after some days.

1. PDA Plate with Malachite Green.

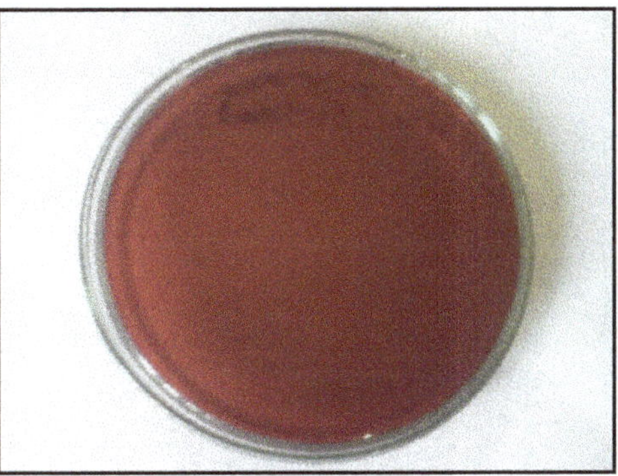

2. PDA Plate with Congo Red.

Zone of Clearance

Observation and Result

Screening of Mushroom Species for Dye Degradation

Table 3.1: A Screening Table for Malachite Green Dye Degradation

S.No.	*Name of Species*	*Observation (Clear zone)*
1.	*Pleurotus florida*	Observed
2.	*Pleurotus sajor-caju*	Observed
3.	*Polyporus* sp. 1	Observed
4.	*Jelly* sp.	Observed
5.	*Schizophyllum commune*	Observed
6.	*Polyporus* sp. 2	Not observed
7.	*Maitake*	Not observed

Screening of Malachite Green Dye by Mushroom

After 5 Days

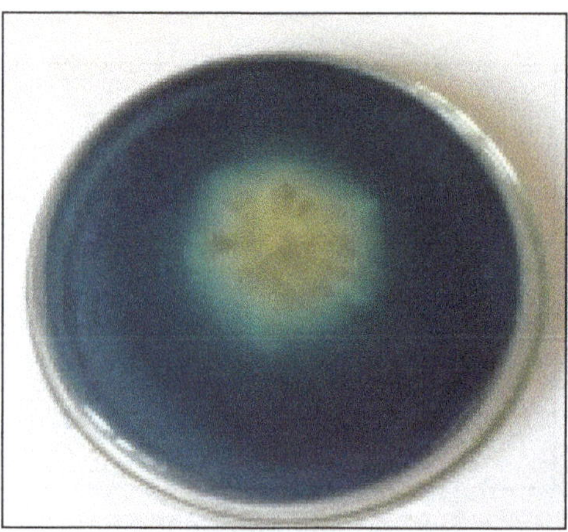

Pleurotus florida

After 10 Days

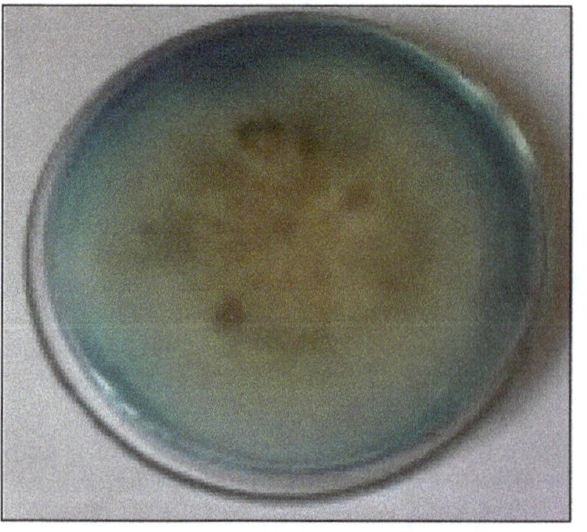

Pleurotus florida

After 5 Days

After 10 Days

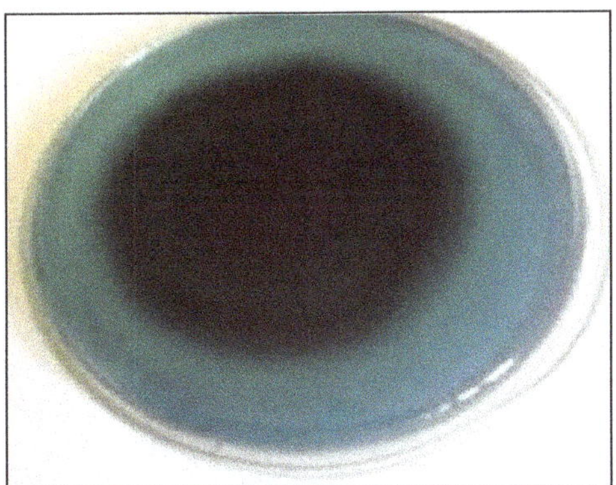
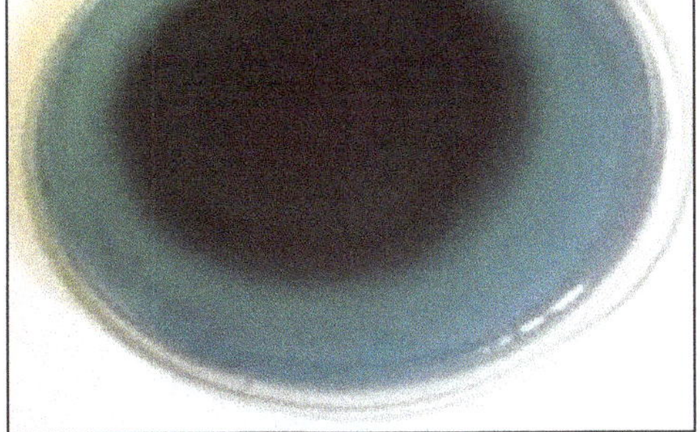

Pleurotus sajor-caju

Pleurotus sajor-caju

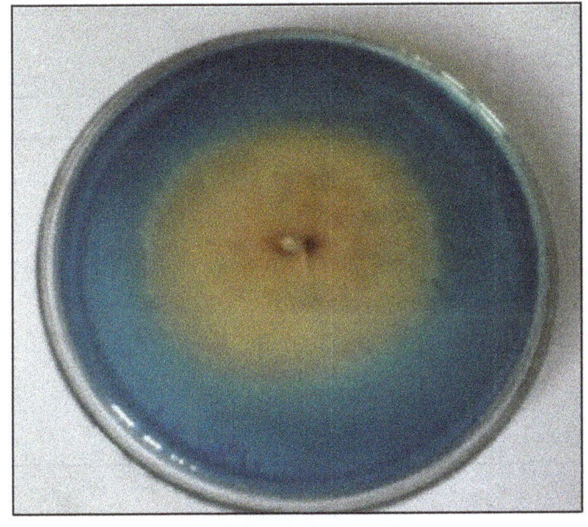

***Polyporus* sp 1**

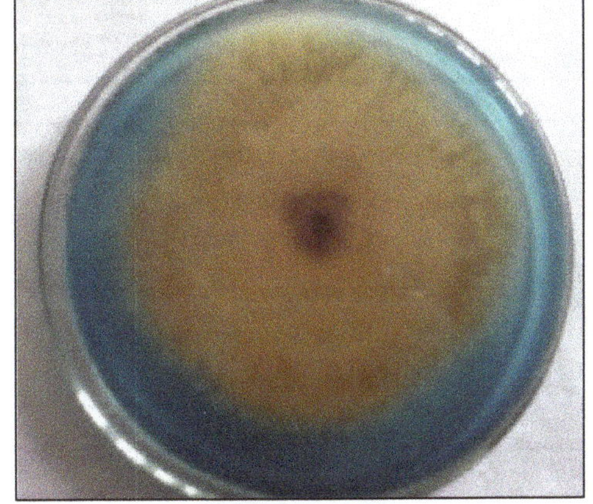

***Polyporus* sp 1**

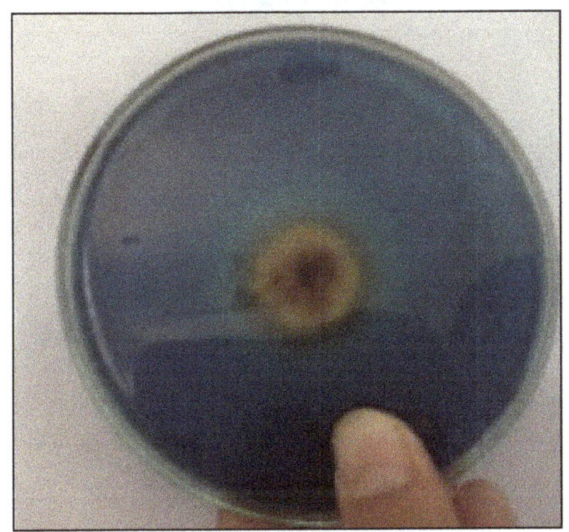

Schizophylum commune

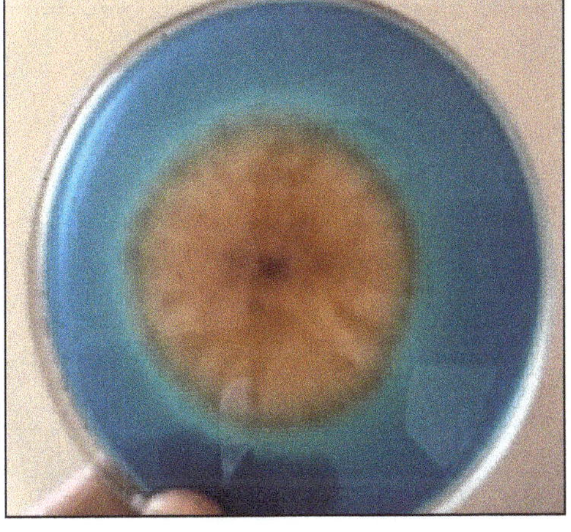

Schizophylum commune

After 5 Days

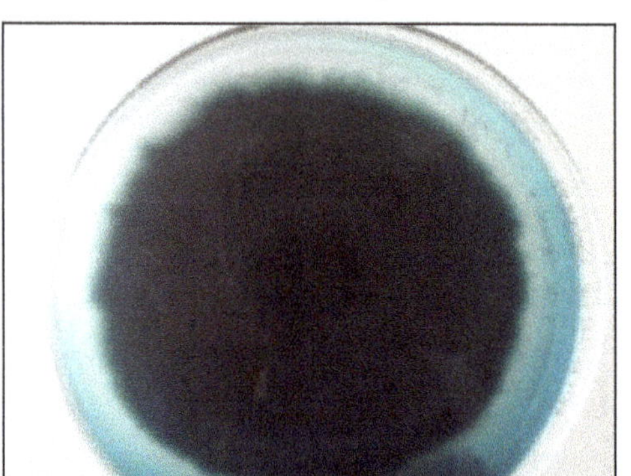

***Jelly* sp.**

After 10 Days

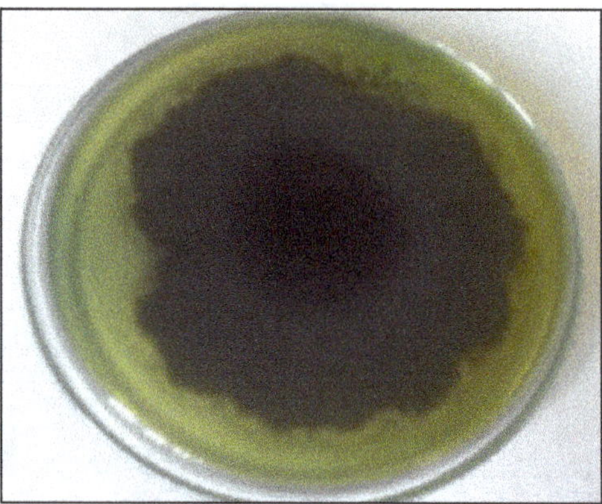

***Jelly* sp.**

Table 3.2: Screening Table for Congo Red Dye Degradation

Sl.No.	*Name of Species*	*Observation (Clear zone)*
1.	*Pleurotus florida*	Not observed
2.	*Pleurotus sajor-caju*	Observed
3.	*Polyporus* sp. 1	Not observed
4.	*Jelly* sp.	Not observed
5.	*Schizophyllum commune*	Not observed
6.	*Polyporus* sp. 2	Not observed
7.	*Maitake* Not observed	

Screening of Congo Red Dye by Mushroom

After 5 Days

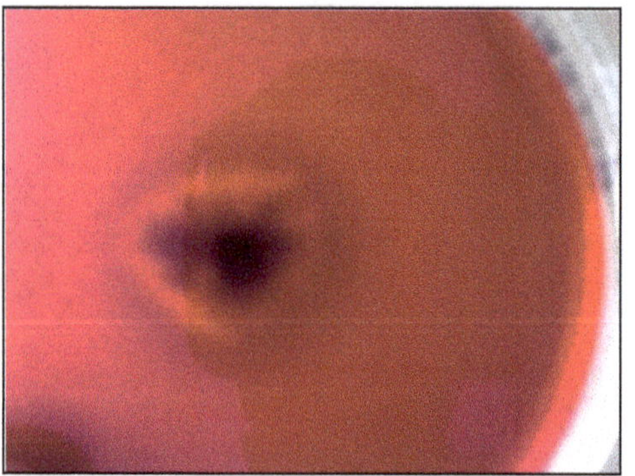

Pleurotus sajor-caju

After 10 Days

Pleurotus sajor-caju

Decolorization percentages

Decolorizing activity was expressed in terms of percentage decolorization and was determined by monitoring the decrease in absorbance at 372 nm of Dis-azo dye (direct brown), against the medium. Decolorization activity (per cent) was calculated according to the formula:

$$\text{Decolorization activity (\%)} = \frac{[(\text{Initial absorbance}) - (\text{Observed absorbance})]}{\text{Initial absorbance}} \; 100$$

Estimation procedure for dye degradation was carried out by the method of Rania M.A. Abedin, 2008.

Potato Dextrose Broth with Dyes (Secondary Screening)

Media Composition

Peeled potato tubers 20, Dextrose 20 g, Dyes 0.001 per cent, Malachite green, Congo red, Distilled water 1000 ml,pH 5.6.

Procedure

500 ml of water in one liter beaker was taken.200 g of washed, peeled and sliced potatoes were added to the beaker. Potatoes were gently boiled for 30 minutes or the time till they were easily penetrated by a glass rod. Filtered through cheese cloth, squeezing out all liquid.20 g dextrose was added to the potato extract.Volume was making up to 1000 ml by the addition of distilled water. Media was added to 50 ml for per duplicate conical flasks and cotton plug was applied.Out of this the one flask was used as a blank and dye was not added. After that rest of the flask was contained dye. All flasks were autoclaved at 121 C, 15 lbs pressure.Flask was allowed to cool until the flask can be held by hand. After that one flask was used as a control, and all other flask in which microorganism were inoculated. All inoculated flask were incubated at 30 C in BOD. After 5 days take first reading, than after 10 days take a second reading by using spectrophotometer.

1. PDB

2. PDB with Malachite Green.

3. PDB with Congo Red.

Flow Chart of Estimation Technique

PDB (Malachite green and Congo red)

Divided into different conical flasks

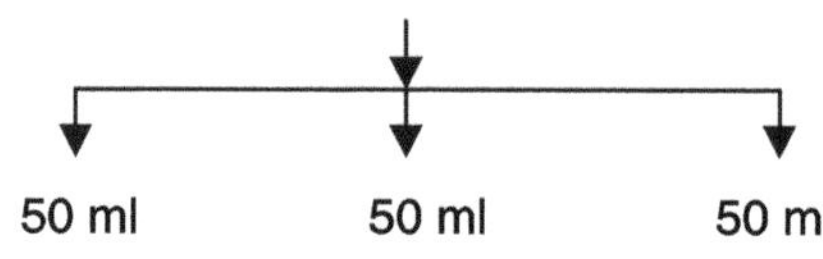

50 ml 50 ml 50 ml

(Blank) (In this added dyes and autoclaved)

One flask used as a control and in all other flasks mushroom were inoculated.

After 5 and 10 days take reading by using Spectrophotometer.

Percentage Degradation of Malachite Green by Mushroom Species

After 5 days

Sl.No.	*Name of Species*	*Remaining Dye (Per cent)*	*Percentage of Dyed Degrade (Per cent)*
1.	*Pleurotus florida*	86.01 per cent	12.99 per cent
2.	*Pleurotus sajor-caju*	85.97 per cent	14.03 per cent
3.	*Polyporus* sp. 1	85.84 per cent	14.16 per cent
4.	*Jelly* sp.	55 per cent	45 per cent
5.	*Schizophyllum commune*	86.2 per cent	13.80 per cent

Percentage Degradation of Malachite Green by Mushroom Species

After 10 days

Sl.No.	*Name of Species*	*Remaining Dye (Per cent)*	*Percentage of Dyed Degrade (Per cent)*
1.	*Pleurotus florida*	47 per cent	53 per cent
2.	*Pleurotus sajo-caju*	52.82 per cent	47.18 per cent
3.	*Polyporus* sp. 1	51.84 per cent	48.16 per cent
4.	*Jelly* sp.	12 per cent	88 per cent
5.	*Schizophyllum commune*	52.07 per cent	47.93 per cent

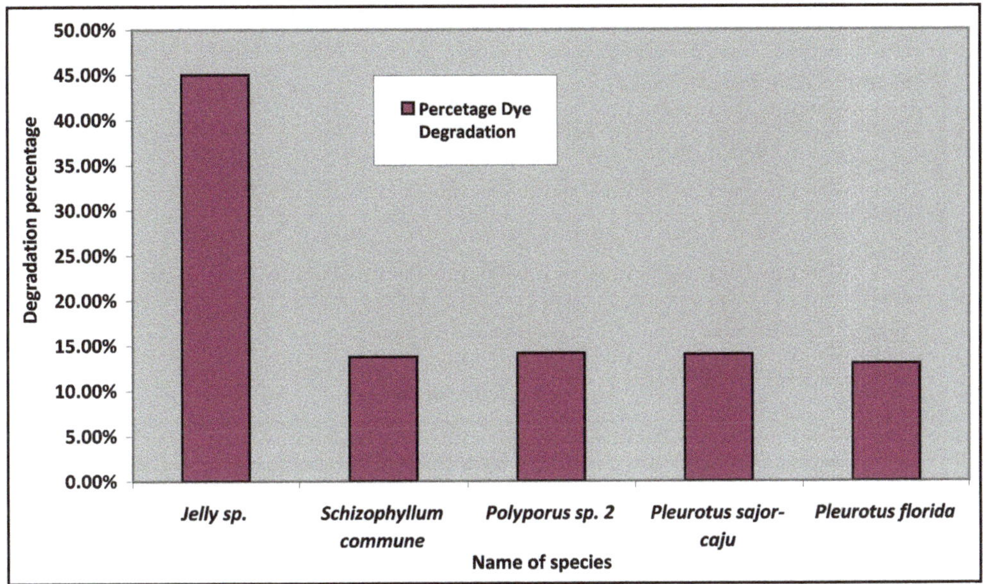

Percentage Degradation of Malachite Green by Mushroom Species after 5 Days.

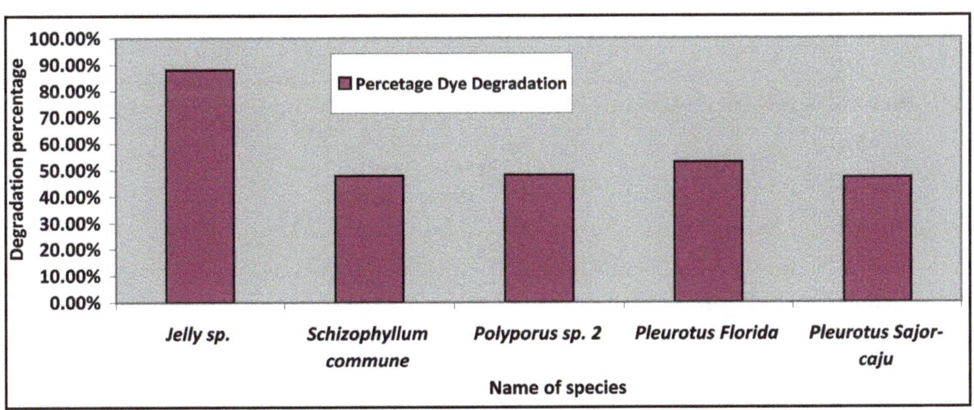

Percentage Degradation of Malachite Green by Mushroom Species after 10 Days.

Percentage Degradation of Cong Red by Mushroom Species

After 5 Days

Sl.No.	*Name of Species*	*Remaining Dye (Per cent)*	*Percentage of Dyed Degrade (Per cent)*
1.	*Pleurotus sajor-caju*	69.5 per cent	30.5 per cent

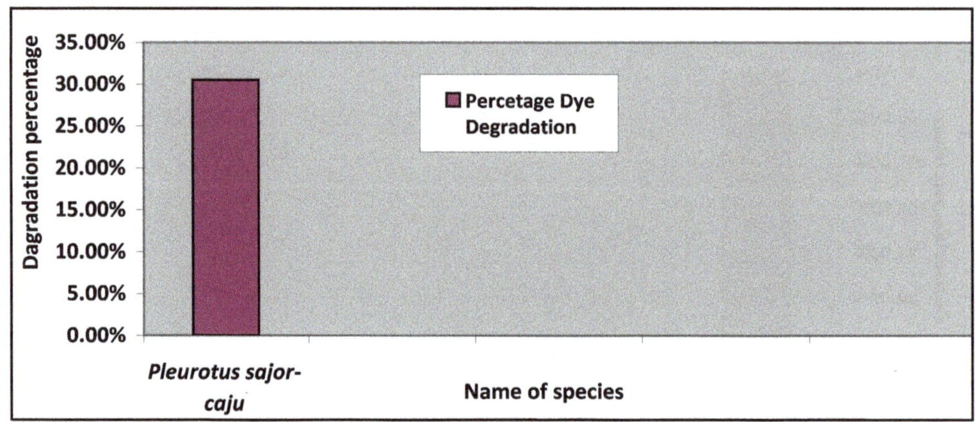

Degradation of Cong Red Dye by Mushroom Species after 5 Days.

Percentage Degradation of Cong Red by Mushroom Species

After 10 Days

Sl.No.	*Name of Species*	*Remaining Dye (Per cent)*	*Percentage of Dyed Degrade (Per cent)*
1.	*Pleurotus sajor-caju*	22.5 per cent	77.5 per cent

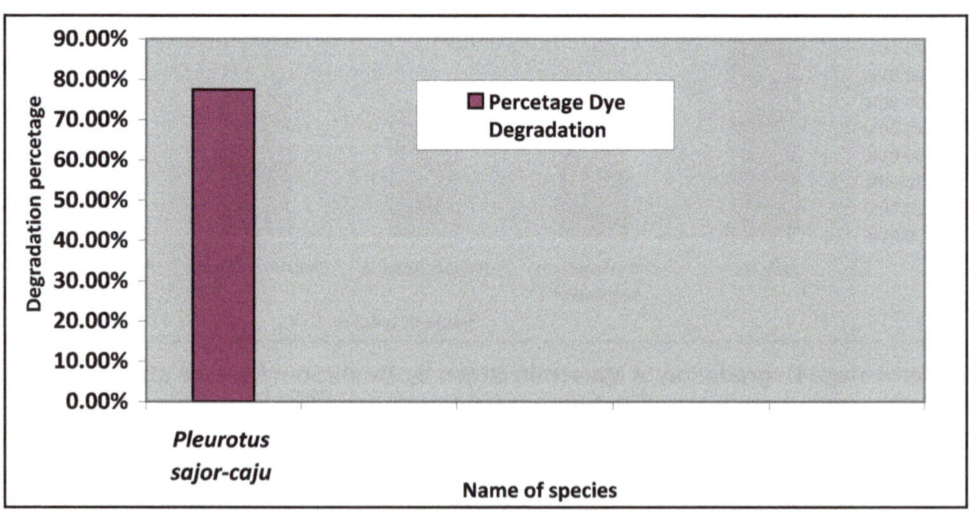

Degradation of Cong Red Dye by Mushroom Species after 10 Days.

Result

Out of seven species *Pleurotus florida, Pleurotus sajor-caju, Polyporus* sp. 1, *Jelly* sp., *Schizophyllum commune, Polyporus* sp. 2 ,*Mitake* only five species *Jelly* sp., *Schizophyllum commune, Polyporus* sp. 2, *Pleurotus florida, Pleurotus sajor-caju* show the dye degradation in Malachite green dye and one species *Pleurotus sajor-caju* shows dye degradation in congo red dye. After 5 days percentage of dye degradation by *Jelly* sp. (45 per cent), *Schizophyllum commune* (13.80 per cent), *Polyporus* sp. 1 (14.16 per cent), *Pleurotus florida* (12.99 per cent), *Pleurotus aajor-caju* (14.03) and after 10 days the same species gives the percentage of dye degradation *i.e.* 88 per cent,47.93 per cent,48.16 per cent, 53 per cent,47.18 per cent. But out of these three species *Jelly sp.* give the high percentage (after 5 days 45 per cent and after 10 days 88 per cent) of dye degradation and in congo red dye after five days percentage of dye degradation by *Pleurotus florida* (30.5 per cent) and after ten days (77.55 per cent).

Conclusion

The elimination of colored effluents in wastewater is based mainly on physical and chemical methods. This is a biological method and biosorption-based process that utilizes the sorption capacity of biological material for the removal of pollutants. using mushroom is the biological method which is best as compared to physical and chemical method for the elimination of colored effluent in wastewater.

References

Aksu, Z., Tezer, S. (2005). Biosorption of reactive dyes on the green alga *Chlorella Vulgaris*. *Process Biochem.* 40, 1347–1361.

Atacag H, Ozyurt M, Taner F (2004). Anaerobic and aerobic consecutive treatments of reactive black 5 azo dye by *Bacillus subtilis*. *Fresen. Environ. Bull.* 13(2): 112-117.

Atacag H, Ozyurt M (2005). Aerobic and anaerobic consecutive treatments of the azo dye reactive black 5 by *Bacillus subtilis*. *Fresen. Environ. Bull.* 14(9): 841-843.

Banat IM, Nigam P, Mc Mullan G, Marchant R, Singh D (1996). Microbial decolorization of textile dye containing effluents. *Biores Technol* 58: 217-27.

B. D. Bhole, Bratati Ganguly, Anusha Madhuram, Deepti Deshpande and Jui Joshi (2004). Biosorption of methyl violet, basic fuchsin and their mixture using dead fungal biomass *Current Science*, Vol. 86, No. 12.

Daneshvar, N., Aleboyeh, A., Khataee, A.R. (2005). The evaluation of electrical energy per order (eeo) for photooxidative decolorization of four textile dye solutions by the kinetic model. *Chemosphere* 59 (6), 761–767.

Deepak Pant and Alok adholeyaidentification (2007). Ligninolytic Enzyme Activity and Decolorization Potential of Two Fungi Isolated from a Distillery Effluent Contaminated Site Springer Science + Business Media B.V.

Erkurt EA, Ünyayar A, Kumbur H (2007). Decolorization of synthetic dyes by white rot fungi, involving laccase enzyme in the process. *Process Biochem.* 42(10): 1429-1435.

Fu, Y., Viraraghavan, T. (2001). Fungal Decolorization of Dye Wastewater, *Bioresourse Technology* 79, 251.

Ganesh Parshetti, Satish Kalme, Ganesh Saratale, and Sanjay Govindwar (2006). Biodegradation of Malachite Green by *Kocuria rosea* MTCC 1532. *Acta Chim. Slov. 53*: 492–498.

Grant, J. And Buchanan, I. (2000). Colour removal from pulp mill effluents using immobilized horse radish peroxidase, SFM Network Project Report.

Gupta, V.K., Mittal, A., Gajbe, V. (2005a). Adsorption and desorption studies of a water soluble dye, Quilnoline Yellow, using waste materials. *J. Colloid Interf. Sci.* 284, 89–98.

Gupta, V.K., Suhas, I.A., Saini, V.K., Garven, V.T., Van Der Bruugen, B., Vandecasteele, C. (2005b). Removal of dyes from wastewater using bottom ash. *Ind. Eng. Chem. Res.* 44, 3655- 3664.

Gurulakshmi. M, Sudarmani. D.N.P and Venba. R. (2008). (Biodegradation of Leather Acid dye by *Bacillus subtilis.*

Karimi, Afzal, Vahabzadeh, Farzaneh, Mohseni, Majid, Mehranian, Mehrnaz (2009). Decolorization of Maxilon-Red by Kissiris Immobilized Phanerochaete Chrysosporium in a Trickle-Bed Bioreactor-Involvement of Ligninolytic enzymesiran. *J. Chem. Chem. Eng. Vol.* 28, No. 2.

Kouichi Nozaki, Chee Hai Beh, Masahiro Mizuno, Tetsuhiro Isobe, Masahiro Shiroishi, Takahisa Kanda, and Yoshihiko Amano (2007). Screening and Investigation of Dye Decolorization.

Luiz Harmed Salmen Espindola; Foued Salmen Espindola; Gláucia Rodrigues de Freitas; Malcon Antonio Manfredi Brandeburgo (2007) Biodegradation of red 40 dye by the Mushroom *Pleurotus* sp., *Florida Biosci. J., Uberlândia, v.* 23, n. 3, p. 90-93,

Mazmanci MA, Unyayar A (2005). Decolourisation of Reactive Black 5 by *Funalia trogii* immobilized on *Luffa cylindrica* sponge. *Process Biochem.* 40: 337–342.

Mcmullan G, Meehan C, Conneely A, Nirby N, Robinson Nigam P, Banat IM, Marchant SWF (2001). Microbial decolourization and degradation of textile dyes *Appl Microbiol Biotecnol* 56: 81-87.

Mehmet A. Mazmanci, Ali Unyayar, Emrah A. Erkurt, Nilay B. Arkçi, Elif Bilen and Mustafa Ozyurt (2009). Colour removal of textile dyes by culture extracts obtained from white rot fungi. *African Journal of Microbiology Research* Vol. 3(10) pp. 585-589.

Mohandass Ramya, Bhaskar Anusha, S. Kalavathy1, and S. Devilaks (2007). Biodecolorization and biodegradation of Reactive Blue by *Aspergillus* sp. *African Journal of Biotechnology* Vol. 6 (12), pp. 1441-1445.

Mohan, S.V., Roa, C.N., Prasad, K.K., Karthikeyan, J. (2002). Treatment of simulated reactive yellow 22 (Azo) dye effluents using Spirogyra species. *Waste Manage.* 22, 575–582.

N. Daneshvar, M. Ayazloo, A.R. Khataee, M. Pourhassan (2006). Biological decolorization of dye solution containing Malachite Green by microalgae *Cosmarium* sp. Elsevier Ltd.

Ozfer Yesilada, Birgul Ozcan (1998). Decolorization of Orange II Dye With the Crude Culture Filtrate of White rot Fungus, Coriolus versicolor, *Tr. J. of Biology* 22 463-476.

Özfer YESILADA Inönü University, Art and Science Faculty, Department of Biology, 44069.Malatya-TURKEY Birgül ÖZCAN Mustafa Kemal University, Art and Science Faculty, Department of Biology, Hatay-TURKEY. Decolorization of Orange II Dye with the Crude Culture Filtrate of White rot Fungus, Coriolus versicolor.

Oxspring, D. A., mcmullan, G., Smyth, W. F. And Marchant, R. (1996). Decolourisation and metabolism of the reactive textile dye, Remazol Black B, by an immobilized microbial consortium. *Biotechnol. Lett.*, 18: 527–530.

Ozyurt M, Atacag H (2003). Biodegradation of azo dyes: a review. *Fresenius Environmental Bulletin (FEB)* 1 (12): 1294–1302

Ozyurt M, Ozer A, Atacag H (2005) Decolourisation of Setopers Black RD-ECO by *Aspergillus oryzae*, Fresen. *Environ. Bull.* 14(6): 531-535.

Ozsoy HD, Unyayar A, Mazmanci MA (2005). Decolourisation of reactive textile dyes Drimarene Blue X3LR and Remazol Brilliant Blue R by *Funalia trogii* ATCC 200800. *Biodegradation* 16: 195–204.

Raina M.A.Botany Department (Microbiology), faculty of science, Alexandria university,Alexandria Egypt. Decolourization and Biodegradation of crystal violet and Malachite green by Fusarium solani saccardo.A comparative study on Biosorption of Dyes By the dead fungal biomass. *American-Eurasian journal of Botany*,1(2):17_13,2008.ISSN 1995-8951.IDOSI publications,2008

Robinson T, Mcmullan G, Marchant R, Nigam P (2001). Remediation of dyes in textile effluent: a critical review on current treatment technologies with a proposed alternative. *Bioresour. Technol.* 77: 247–255

Unyayar A, Mazmanci MA, Atacag H, Erkurt EA, Coral G (2005). A drimaren blue X3LR dye decolorizing enzyme from *Funalia trogii*: one step isolation and identification. *Enzy. Microbial Technol.* 36: 10–16.

Verma P, Baldrian P, Nerud F (2003). Decolourization of structurally different synthetic dyes using cobalt (II)/ascorbic acid/hydrogen peroxide system. *Chemosphere* 50: 975-979.

Wafaa M. Abd El-Rahim1*, Ola Ahmed M. El-Ardy2 and Hassan Moawad1, (January 2008).Aeration as a factor in textile dye bioremoval by *Aspergillus niger, African Journal of Biochemistry Research* Vol.2 (1), pp. 030-039,

Wafaa M Abd-El Rahim, Moawad H (2003).Enhancing bioremoval of textile dyes by eight fungal strains from media supplemented with gelatin wastes and sucrose. *J. Basic Microbial.* 43(5): 367-375.

Wafaa M Abd-El Rahim, Moawad H, Khalafallah MA (2003).Microflora involved in textile dye waste removal. *J. Basic Microbiol.* 43(3): 167-174.

Wesenberg D, Kyriakides I, Agathos SN (2003).White-rot fungi and their enzymes for the treatment of industrial dye effluents. *Biotechnology*. Adv. 22 161–187.

Zhang F, Yediler A, Liang X, Kettrup A (2004).Effects of dye additives on the ozonation process and oxidation by products: a compartive study using hydrolysed CI Reactive red 120. *Dyes Pigments* 60: I-7.

Pesticides: 'A Threat to Mankind'

Madhurani Shukla and K.K. Tiwari

Department of Chemistry,
Govt. Nagarjuna P.G. College of Science, Raipur

ABSTRACT

Pesticides are chemical or biological agents mainly used in agricultural production to prevent pests, weeds, diseases and other plant pathogens from damaging the crops and reduce yield losses. Besides these, pesticides are also used in homes, school, forests and roads. Our world is almost filled with pesticides.

Pesticide formulations contain both 'active' and 'inert' ingredients. Active ingredients kill the pests and inert ingredients help the active ingredients to work more effectively. Solvents used as an inert ingredient in many pesticide formulations may prove to be toxic if inhaled or absorbed by the skin. The pesticides have adverse effects on the environment due to their toxicity. Consumption of pesticides coated agricultural products result into food poisoning. Pesticide poisoning may cause symptoms like nausea, abdominal cramps and anxiety. Due to the health hazards it is essential to assess the impact of pesticides on human health. But such assessment is not an easy and accurate process as periods and exposure level to pesticides, their types may differ and the environmental conditions where pesticides are generally applied also vary.

Pesticides are toxic and they have harmful effects on organisms other than pests including human. Particularly children are more vulnerable to the toxic effect of pesticides to an extent that even very low levels of exposure during development may have adverse health effects. Risk of pesticide on human health can be assessed by a number of criteria which help to assess the adverse effects of them and their implementation. These would also possibly effect the characterization of already approved pesticides and will also affect the approval of new compounds in future. Hence new sensitive, reliable methods are needed to predict hazardous nature of pesticides and thus contribute to reduction of the adverse effects on human health and environment.

Keywords: *Pesticides, Health of human, Control various pests, Pesticides risk.*

Introduction

Pesticides are naturally occurring and man-made substances which are used for controlling various pests such as insects, weeds and plant pathogens. The term pesticide refers to chemical substances that are very active and interfere the biological process of living organisms. Most of the pesticides are slowly degradable and thus remain in the environment causing harmful effects on animals, plants and human beings (1).

Pesticides are used to manage various pests in agriculture But pesticides are also responsible for the death of a large number of human beings and animals.(2) In this topic we focus on the worldwide use of pesticides in agriculture and compare the benefits and problem associated with very important but very dangerous chemicals (8).

Use of Pesticides

Agriculture was developed to give crops and foods for human and other animals. Human population increased rapidly but the agricultural area where we produced food and crops are inadequate. So the amount of food produced is very precious. Unfortunately, agricultural pests destroy nearly 33 per cent of all crops produced, which results in economic losses (5).

Due to this factor pesticides are often used to try to remove the problem. Pesticides are chemicals that kill or control the population of pests. There are many kinds of pesticides like herbicides, insecticides, rodenticides fungicides algaecides etc. but the common pesticides are herbicides and insecticides which kill unwanted plants and insects. The use of pesticides increased rapidly which had several benefits and also caused many problems.

Analysis of Pesticides

The detection and quantification limits should be reliable and applicable to a range of analytical methods. The important question to risk evaluation is assessing the risk to a human being from consuming various foods treated with pesticides. When a pesticide having residue with no detectable range is registered for use "Tolerance" is recognized at the lowest concentration range at which the method was validate.(11,12) But this would be assume that all food products treated with pesticide contain some amount of pesticide and for which detectable residue were not found is just impossible then the lowest level of method validation(13,14).

Benefits of Pesticides

According to the use of pesticides it is possible to destroy pests and increase the production of food. And therefore the farmers are able to increase their profit by selling their production. Using pesticides saves time and money as compared to removing weeds and pests from fields manually. In addition to saving crops and foods, pesticides have also given direct benefits to human health. Use of pesticides was also beneficial for man as they killed pests which carry or transmitted various diseases Malaria, plague etc. were minimized due to the use of pesticides.

Pesticides are used in our homes and businesses for example plastics and various paints may possibly contain fungicides to prevent moulds. Toilet cleaners and disinfectants also contain pesticides. Packaged grocery products are protected from insect contamination by the controlled use of pesticide *i.e.* insecticides in manufacturing and packaging process (6). Pesticides manage insects as well as rodents. It is obvious that proper use of pesticides not only improves our life and protects our property but also gives better environment.

Problems Associated with Pesticides

Use of pesticides not only gives several benefits but also have many problems associated with their use. When pesticides are used they remain in the environment and often move through air, water and soil. Although pesticides are beneficial yet they also pose risk. The hazards of pesticides are not the only problems but the question is to whom and to what degree they pose these risks.

Effect through Food Commodities

In India, poisoning due to pesticides was first reported in Kerala in 1958 where about 100 people died after consuming wheat flour contaminated with parathion to assess the pesticide residues selected food commodities collected from various part of country. DDT residues were found to be above the tolerance limit (9).

Effect on Environment

Pesticides can be present in soil, water and other vegetation. Along with killing insects or weeds, pesticides also show toxic effect to a host of other organism including fish, birds and non target plants.

Effect on Ecosystem

Pesticides also disrupt the balance of an ecosystem. When a pesticide is used it also kills non pest organism. The environment can be changed to make it favorable for the pest. Human exposure to pesticides has caused poisoning leading to development of deadly diseases like Cancer (16). Generally pesticides are not broken down and therefore when they are consumed, they are permanently stored in the body.

Contamination in Surface Water

Pesticides also reach the surface water through runoff water from agricultural field. It was found that more than 90 percent of runoff water contained several pesticides.

Contamination in Groundwater

Pollution of groundwater from pesticide is a major global problem. According to one survey in India about 58 per cent of water samples from wells and hand pumps around Bhopal were contaminated with organo-chlorine pesticides (7). Once the water is polluted with pesticides, they take a long period to clean up this contamination. These processes are very complex as well as costly.

Contamination in Soil

A wide range of pesticides have been found in a various transformation products (TPs) which have been monitored in soil.(10) It was found that TPs in soil are generally obtained from herbicides. The amount of pesticides and their TPs present in soil depend on the interactions between the soil and the pesticides properties.

Effect on Soil Fertility

Soil microorganisms which are very beneficial can be decline due to heavy treatment of soil with pesticides.

Pesticides and Kids

Pesticides residues give an intolerable risk to children. Dangerous levels of pesticide residues in fruits and vegetables may lead to cancer or neurologic disorder in children, the most terrible being some of the foods that kids like best *i.e.* banana, orange, fruit juices and sweets like jalebis and toffees etc (4).

Conculsion

Pesticides are used for a quick, easy and low expensive method for controlling weeds and insect pests. Pesticides have contaminated every component of our environment. Pesticides residues are present in air, soil, surface water and ground water around the countries. Pesticides contamination poses risk to environment, to insects, birds and plants.

benefits and national benefits. It is necessary to convey the message that preventing the adverse health effects and progress of health are profitable investments for employers. It is thus the reason to develop health education packages to minimize human exposure from pesticides (15).

Acknowledgements

The authors are thankful to the Principal Dr. S. V. K. Prasad, Head of Department Dr. Shipra Verma, Govt. N.P.G. College of Science, Raipur, for providing laboratory facilities. One of the authors is thankful to Dr. (Mrs.) Vinod Sharma Principal, Govt. College, Gobra Navapara for their cooperation.

References

1. F.A. Gunther and L.R. Jeppson, (1960). "Modern insecticides and world food production", Champan and Hall, London.
2. N.Dudley, (1987). "This Poisoned Earth", Platkus publishers Ltd. London.
3. L. Robert, (1989) *Science*, 243, 1280.
4. Christos A. Damalas and Llias G. Eleftherohorinos, (2011). "Pesticide Exposure, Safety issues and Risk assessment indicator" *Int J Environ Res Public Health*, 8(5), 1402-1419.
5. Christos A. and Damalas, (2009). "Understanding benefits and risks of pesticide use" *Science Research and Essay* Vol.4 (10), 945-949.

6. Davis J.R., Brownson R.C. and Garcia R, (1992). "Family pesticide use in the home, garden, orchard and yard" *Arch. Environ. Contam. Toxicol.* 22, 260-266.
7. Fishel F.M. (2007). "Pesticide use trend in the U.S.: global comparison. Florida Cooperative Extension Service, University of Florida. P.-143.
8. Metcalf R.L. (1987). " Benefit/risk considerations in the use of pesticides."Agric *Human values* 4, 15-25.
9. Vander Werf H.M.G. (1996). "Assessing the impact of pesticides on the environment". *Agric Ecosyst. Environ.* 60, 81-96.
10. Woodruff T.J., Kyle A.D. and Bois F.Y. (1994). "Evaluating health risks from occupational exposure to pesticides and the regulatory response." *Environ. Health Perspect.* 102, 1088-1096.
11. Charizopoulos E. and Morikidou E. P. (1999). "Occurrence of pesticides in Rain of the Axios River Basin,Greece", *Environ. Sci. Technol.*,33,2363-2368.
12. Costa L.G. (1987). "Toxicology of Pesticides: Experimental, Clinical and Regulatory Perspectives", Springer-Verlog, Berlin.
13. Long G. L. and Winefordner J. D. (1983). *Anal. Chem.* 55,713A.
14. Miller J.C. and Miller J.N. (1988). "Statistics for analytical Chemistry" Ellis Horwood Chichester.
15. Masti S.P., Seetharamappa J., Melwanki M.B. and Motohashi N.,(2002). "*Anal. Sci.*", !8, 167-169.
16. National Toxicology Program, (2011). " Report on Carcinogens Twelfth Edition.

Chlorococcales Algae in Narsaiya Pond of Raipur City

***Smriti Chakravarty*[1] *and M.L. Naik*[2]**

[1]*Department of Botany, Govt. Chhattisgarh College,*
Raipur – 492 001, Chhattisgarh
[2]*Retired Professor, SOS in Bioscience,*
Pt. Ravishankar Shukla University, Raipur, Chhattisgarh

ABSTRACT

A taxonomic survey of the Chlorococcales algae in Narsaiya pond of Raipur city was carried out for one year. Samples were gathered monthly during the period of January to December. Seasonal variation observed in the density of chlorococcalean algae in the pond appear to be due to two factors: one being the temperature, the seasonal variation of which had caused the variation, season wise. The other factor had been the growth of cyanobacteria. Increase in the density of cyanobacteria, during summer months has always caused the decline in the chlorococcalean density. In Narsaiya pond 10 forms were recorded.

Keywords: *Chlorococcales, Seasonal variation, Raipur city.*

Introduction

Raipur city is the capital of newly constituted Chhattisgarh State. Within the Raipur city area pond was selected as study sites. Pond of Raipur city area is receiving huge amounts of sewage and sullage. Polluted water of this pond is supporting several groups of algae. One of the important group, with respect to density, is chlorococcales. Narsaiya pond is situated in the southern part of the city on Dhamtari Road. The pond covers an area of about 10.1 hectares. Owing to availability of ample water, a slum of large number of families have settled around this pond, on its bunds. The water is being mostly used by the slum dwellers mostly for *nistar* purposes but to some extent for drinking purpose also.

Materials and Methods

Samples were collected in plastic cans. Before collecting the sample, the cans were washed thoroughly with tap water and then were rinsed several times with the pond water to be collected. Surface water only was collected from the pond, for experimentation. To identify the chlorococcalean algae, from the pond water samples were collected in the following manner: A vial of about 50 ml was tied to the bottom of plankton net. Through the net 10 litre of surface water was filtered from the pond. Lugol's iodine solution was added as preservative to this sample. Volume of this vial-collected sample was made up to 100 ml. These preserved samples were observed after words for the presence of chlorococcalean algae at convenient time.

Results

In Narsaiya pond 10 forms were recorded. *Tetraedrom minimum* present only in 2 samples was the most rare species, it was detected only the month of may and July. *Scenedesmum armatus* var. *bicaudatus, Scenedesmus quadricauda, Scenedesmus bijugatus, Scenedesmus quadricauda, Schroederia indica* were present throughout the year and also the most dominant forms of chlorococcalean algae in this pond. *Chlorococcum infusionum* could not be recorded in the months of April and June. *Micractinium pusillum* was present only in 7 of the months, being absent from April to July and October. *Pediastrum ovatum* was also present only in 7 of the months and absent in the month of April, June, July, October and November. *Scenedesmus armatus* var. *major* was little more regular in its occurrence, was present in 10 months and absent only in April and June. *Tetraedron muticum* was present for 8 of the months with its absence from the month of April to the month of July (Table 5.1).

1. ***Chlorococcum infusionum*** (Schrank) Meheghini Philipose, 1967, p.73, fig- 1 —Cells 10-109µ in diameter chloroplast like a hollow spherical with a single pyrenoid.
2. ***Micractinium pusillum*** Fresenius Philipose, 1967, p-104, fig-29 — Colonies quadrate with 4-16 cells arranged in group of four. Cells spherical and 3-10µ diameter. Chloroplast single, parietal, cup-shaped and with a pyrenoid.
3. ***Pediastrum ovatum*** (Ehr.) A. Braun Philipose,1967, p-115, fig-37 – Colonies 4-8 celled with the cells arranged in a ring round a central space, four celled colonies 60µ and eight-celled colonies 80µ in diameter.
4. ***Scenedesmus armatus* var. *bicaudatus*** (Gulielmetti) Chodat, Philipose, 1967, p-262, fig-171 – Colonies two or four celled, the spines of the two terminal cells alternating with each other.
5. ***Scenedesmus armatus* var. *major*** G. M. Smith Philipose, 1967, p-266, fig-171 – Colonies four celled, cells oblong ellipsoid with acute spices and arranged in a linear series, terminal cells with a single long spine from each pole. Cells 8.8 – 9.7µ broad, 24.6- 26.4µ long.Spines 15-17.6 µ long.
6. ***Scenedesmus bijugatus*** (Turpin) Kuetzing Philipose, 1967, p-252, fig-164 – Colonies flat, of 2-4 cells arranged in a single linear series. Cells oblong-ellipsoid with the ends broadly rounded. Cells 3.5 - 7µ broad, 7 - 23µ long.

Table 5.1: Seasonal Variation in the Density of Chlorococcalean Algae in Narsaiya Pond

Algae	*Jan*	*Feb*	*Mar*	*Apr*	*May*	*Jun*	*July*	*Aug*	*Sep*	*Oct*	*Nov*	*Dec*
Chlorococcum infsionum	61.36	27.00	22.08	—	12.26	—	7.36	56.42	27.00	24.54	39.24	34.36
Micractinium pusillum	14.72	9.82	7.36	—	—	—	—	12.26	12.26	—	12.26	14.72
Pediastrum ovatum	22.08	4.90	4.90	—	2.46	—	—	14.72	14.72	—	—	17.18
Scenedesmum armatus var. *bicaudatus*	61.36	53.98	44.18	181.64	159.56	184.10	184.10	181.64	53.98	44.18	46.64	41.70
S.armatus var. *major*	4.09	19.62	12.26	—	2.46	—	2.46	7.36	7.36	4.90	4.90	7.36
Scenedesmusbijugatus	41.70	66.28	51.54	201.28	196.38	191.46	193.92	196.38	71.18	66.28	56.42	56.42
Scenedesmus quadricauda	71.18	53.98	73.64	61.36	83.46	51.54	51.54	73.64	71.18	83.46	61.36	63.82
Schroederia indica	27.00	39.24	51.54	4.90	4.90	7.36	7.36	24.54	51.54	31.90	29.46	29.46
Tetraedronminimum	—	—	—	—	2.46	—	4.90	—	—	—	—	—
Tetraedron muticum	19.62	31.90	17.18	—	—	—	—	14.72	14.72	12.26	14.72	17.18

7. ***Scenedesmus quadricauda*** (Turpin) Brebisson Philipose, 1967, p-286, fig-187 – Colonies 4 celled, cells oblong- cylindrical with rounded ends and arranged in a linear series, terminal cells with a long, curved spine, cells 3-7μ broad, 9 - 18μ long, spines 6.5 - 15μ long.

8. ***Schroederia indica*** sp. nov. Philipose, 1967, p-90, fig-19 – Cells more or less semicircular, chloroplast parietal and one in number. Cells 4.4 - 12.3μ broad and 68 -84μ long with spines.

9. ***Tetraedron minimum*** (A. Braun) Hansgirg. Philipose, 1967, p-138, fig-53 – Cells crescent shaped, curved and pointed ends, solitary, cells 2 – 3μ broad, 7-9μ long.

10. ***Tetraedron muticum*** (A. Braun) Hansgirg. Philipose, 1967, p-137, fig-51 – Cells small, flat and triangular with the sides slightly concave and angles broadly rounded. Cells 6-30μ in diameter.

References

Bharati, S.G. 1964. Chlorococcales from Kodai Kanal, South India. *J. Bombay Nat. Hist. Soc.* 61: 474-479.

Chaturvedi, U. K. and I. Habib 1996. A systematic account of Chlorococcales from Nepal. *Phycos.* 25: 129-137.

Iqbal Habib and Chaturvedi, U.K. 2002. A systematic account of Chlorococcales from Mahoba, India. *Phycos* 40 (1 and 2): 107-113.

Laal, A.K. 1976. A check list of Chlorococcales from fish ponds of Patna. *Phycos* 15: 101-102.

Mathur, M. and Pathak, V.1990. Some Chlorococcales new to India. *Phykos*.29: 111-113.

Philipose, M.T. 1967. *Chlorococcales*, ICAR. New Delhi.

Sinha, S. and Naik, M.L. 1997. *Phytoplankton and Macrophytes in the ponds of Raipur city area.* Pt. R. S. University Raipur.

Tarar, J.L. and Bodhke, S. 1998. Studies on Chlorococcales of Nagpur. *Phykos.* 37 (1-2): 107-114.

6

Study of Airspora in Different Vegetable Market in Raipur City with Special Reference to Winter Season

Priti Tiwari

Principal, Govt. R.L. College, Rajim

ABSTRACT

Airspora constitutes fungal spores, pollen grains, bacteria, hyphal fragments, insect's scales etc. Fungal spores constitutes a major components of airspora. Fungal spores are part and parcel of air and their quality and quantity depends on geography, seasonal variation in local environment.

Fungi are ubiquitous in nature and they are present in both indoor and outdoor environments. The fungal spores remain suspended for longer time in the air, their presence depend on the various factors like temperature relative humidity, rain fall, seasonal variation etc. Suspension of organic and inorganic material also effect the distribution of microbes in the air. The present study was aimed to investigate airspora of vegetable markets. The study was conducted from 7 different vegetable market at Raipur city. The sampling was done by gravity petri plate exposure technique with PDA medium. Total 19 fungal types and 6 bacterial forms were recorded during the study period. The frequently isolated fungi were Rhizopus, Aspergillus, Alternaria, Cladosporium, Penicillium and Fusarium. The most dominant fungal species of Rhizopus nigricans Aspergillus flavus, A. niger, Penicillium species. The present work was carried out on different vegetable market to study fungal diversity. The bacteria and fungal propagules were the major components in the study area.

Keywords: *Fungal diversity, Airspora, Vegetable markets, Aspergillus, Bacteria.*

Introduction

Fungi are universally present in all types of natural habitats and formm one of the most important component of an eco system as decomposers. It has been known that vegetables and fruits plays a vital role in human nutrition by supplying the necessary growth factors such as vitamins and essential minerals in human daily diet and that can help to keep a good and normal health. Fungi play an important role for deterioration of vegetables. Fruits and vegetables market are know to contain several species of fungi.

Environmental aero mycology contains one of the major aspects, mainly because of the dominants of fungal spores in the airspora (Tilak, 1991). The spores are obtain liberated in the air in massive concentration and can remain air born for a long time. The study of aerobiology has its bearing on various aspects of human health and welfare. The present study was carried out to identify fungal forms in different vegetable market in Raipur city and to study their variation in concentration. The study of airspora of market may have some implications on the health of the people working in the market, customers, sellers etc. Fungal propagules in the ambient air is regularly can continuously inhaled by human beings. Consequently it is understandable that the high concentration of most of the fungal spores present in the vegetable market environment.

Air born fungi are considered to act as indicator of the level of atmospheric bio pollution. The presence of fungal spores volatiles and myco toxins in the air can causes health hazards in all segments of the population (Kakde *et al.*, 2001).

Vegetable market is a common place where everybody use to go. As Raipur is a big place it has yet many small and big vegetable market which remains crowdy day and night. So in this survey of various microorganisms present in the air has been given. It includes comparative study of microbes found in vegetable market of Raipur city with special reference to winter season.

Materials and Methods

The survey of airspora was undertaken in the vegetable market at different localities of Raipur the experimental work was carried out continuously for winter season. The culture plate exposure method was used for trapping the airspora. Potato dextrose streptomycin agar was used as culture medium. These Petri plates were exposed for five minutes, The Petri plates were incubated in the laboratory for 7 days at 25 $\pm 1^0$C at the end of incubation period fungal colonies were counted, isolated and identified with the help of available literature. (Barnett and Hunter, 1972 and Subramaniam 1971, Nigmani *et al.*, 2006). The percentage frequency and percentage contribution of airspora also calculated as per standard formula (Jadhav and Tiwari, 1994).

$$\text{Per cent Frequency} = \frac{\text{Number of observation in which species appeared}}{\text{Total number of observation}} \quad 100$$

$$\text{Per cent Contribution} = \frac{\text{Total number of colonies of a species in all observation taken together}}{\text{Total number of colonies}} \quad 100$$

Results and Discussion

The study of airspora indicates 6 types of bacterial colonies and 19 type of fungal species. Fungal species belongs to 8 genara, in which 02 species belongs to Zygomycotina, 12 belongs to Ascomycotina, 05 belongs to Anamorphic Fungi. Maximum no. of fungal colonies 173 were isolated from Kota vegetable market while minimum no of colonies *i.e.* 40 were isolated from Amapara market. Total 557 fungal and 637 bacterial colonies were isolated from different vegetable market in Raipur city.

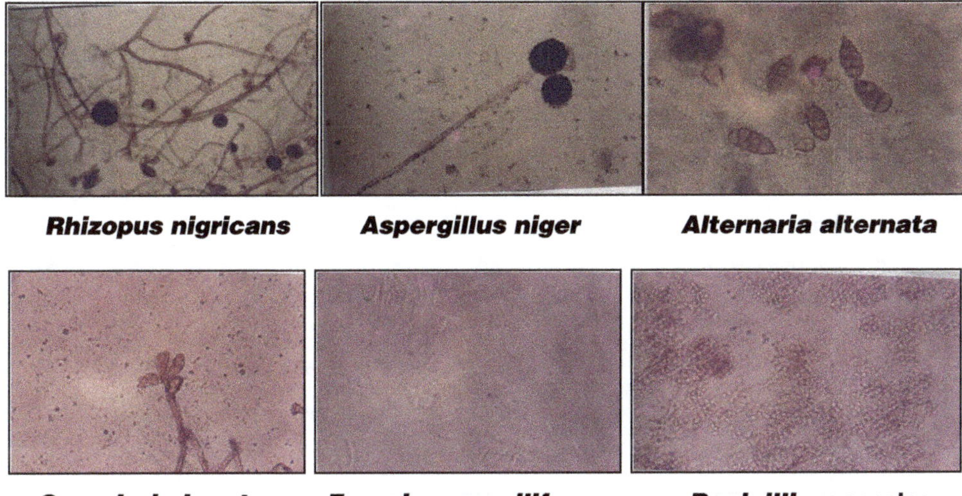

Rhizopus nigricans ***Aspergillus niger*** ***Alternaria alternata***

Curvularia lunata ***Fusarium moniliforme*** ***Penicillium* species**

Table 6.1: Study of Airspora in different Vegetable Market at Raipur city in Winter Season

Sl.No.	*Name of Micro-flora*	*Amapara*	*Dang-aniya*	*Mohba Bazar*	*Kota*	*Shastri Bazar*	*Station Road*	*Tati-bandh*	*Total Colonies*
Zygomycotina									
1.	*Mucor stolonifer*	–	–	–	4	–	–	–	4
2.	*Rhizopus nigricans*	2	1	2	99	3	1	1	109
Ascomycotina									
3.	*Aspergillus flavus*	4	2	–	1	2	–	3	12
4.	*A. fumigatus*	7	5	16	16	8	–	5	57
5.	*A. luchensis*	–	–	1	–	–	–	1	2
6.	*A. nidulans*	10	–	30	2	–	–	11	53
7.	*A. niger*	14	5	24	41	24	4	39	151
8.	*A. parasiticus*	–	–	–	–	–	40	–	40
9.	*A. sulphureus*	–	–	–	4	–	–	–	4
10.	*A. terreus*	–	–	4	1	–	–	1	6

Contd...

Table 6.1–*Contd...*

Sl.No.	*Name of Micro-flora*	*Amapara*	*Dang-aniya*	*Mohba Bazar*	*Kota*	*Shastri Bazar*	*Station Road*	*Tati-bandh*	*Total Colonies*
11.	*A. speluneus*	3	–	–	–	1	–	–	4
12.	*Penicillium funiculoum*	–	–	–	–	4	–	–	4
13.	*P.* species	–	1	41	2	–	–	4	48
14.	*Nigrospora* species	–	–	1	–	–	–	–	1
Anamorphic Fungi									
15.	*Alternaria alternata*	–	2	–	–	–	–	–	2
16.	*A. solani*					2			2
17.	*Cladosporium cladosporioides*		23	9	1			2	35
18.	*Fusarium moniliforme*	–	15	–	2	–	–	1	18
19.	*F. pallidoroseum*	–	2	–	–	3	–	–	5
Bacterial Forms									
1.	Bact. orange colony	–	–	4	–	–	–	–	4
2.	Bact. yellow light	–	1	–	45	2	–	398	446
3.	Bact. pink	–	1	–	21	–	–	5	27
4.	Bact. yellow	–	29	–	–	2	–	–	31
5.	Bact. white	–	15	–	–	–	–	112	127
6.	Bacteria undulating colony	–	–	–	–	2	–	–	2
	Total	**40**	**102**	**132**	**239**	**53**	**45**	**583**	**1194**

During present study some fungal spores as *Aspergillus niger* (151 colonies), *Rhizopus nigricans* (109), *A. fumigatus* (57), *A. nidulans* (53), *A. parasiticus* (40), *Penicillium* species (40), *Cladosporium cladosporioides* (35) were dominant and abundant in all vegetable market. The result of the present study showed a number of similarities to similar surveys conducted elsewhere. This is an agreement with earlier workers Ahmed *et al.* (1991) at Faisalabad, Sahney and Purwar (2002) at Allahabad, Sharma *et al.* (2004) at Silcher, Assam, Rijukakati *et al.* (2007) at Assam, Sonia *et al.* (2009) at Bilaspur, Ahire and Sangale (2012) at Pune, Massoud (2013) at Aswan Egypt, Raveesha (2015) at Kolar, Karnataka.

The study provides the over view of the different fungal population in and around in vegetable markets and to bring awareness to the people about air borne fungal spores and their spoilage in fruits and vegetables and also various allergic disorders caused by these fungus in human beings.

References

Ahire, Y.R. and Sangale, M.K. (2012) Survey of aeromycoflora present in vegetable and fruit market. *Elixir Appl. Botany* 52 (2012) 11381-11383.

Barnett H. L. and Hunter H.B. (1972). Illustrated genera of imperfect fungi, Burgess, publishing, Company, Minneapolis, Minnesota 121.

Dhruba Sharma, B.K Datta and A.B. Singh (2004). Studies on seasonal and annual variations in atmospheric pollen and fungal spores of greater Silchar Assam Ind. *J. of Aerob.* Vol 17 Nos. I and II PP I-II.

Jadhav, S.K. and Tiwari, K.L. (1994). Aeromycoflora of Ravan village Ind. *Bot. Report* 13 (1+2) p-33-36.

Kakade H.U. and Saoji A. A. 2001. Seasonal variation of fungal propagules in a fruit market environment Nagpur India. *Aerobiologia,* 17:177-182.

Kakati, R., Baishya, R. and Sharma G.C. (2007) Study of the aeromycofloral population of west Guwahati. 14th Nat. Conf. Aerob. Raipur, P-37.

Mohamed S Massoud (2013). Survey of fungal disease of some vegetables and fruits in Aswan, Egypt. IOSR-JPBS, vol 6 Issue – 3 (May- June 2013) pp. 39-42.

Nagamani A., Kumar I. K. and Manoharachary C., (2006). Handbook of soil fungi I. K. International Pvt. Ltd. New Delhi India.

Nazir A. Mukhtar, A., M. Aslam Khan, S.T. Sahi and M.N, Bajwa (1991). Studies on the fungi causing vegetable rot in the market and cold storage. *Pak.J. Agri. Sci.* Vol. 28 No.3.

Raveesha, H.R. (Feb 2015). Aeromycolgical study of fruit, vegetable market of Kolar district, Karnataka.

Sahney M. and A. Purwar (2002). Incidence of fungal airspora in the market area of Allahabad. *Ind. J. Aerob.* Vol.15 No. 1 and 2 pp.32-46.

Sonia, A., A. Kabir and U. Singh (2009). A year round aero – mycological study by culture plate technique at Govt. P.G. J.P. Verma Arts and Commerce College Bilaspur (C.G.) NSA FPS Bilaspur p-97.

Subramaniam C.B. 1971. Hyphomycetes, ICMR, New Delhi.

Tilak S.T. 1991.Fungi and biotechnology. Today and Tomorrow Publisher, New Delhi India 137.

7

Ethnomedicinal Use of some Weeds on Rice Fields of Rajim Region, Distt. Gariaband, Chhattisgarh

***Roopshikha Agrawal*[1] *and Arvind Agrawal*[2]**

[1]*Department of Botany,*
Govt. Rajeevlochan College, Rajim, Distt. Gariaband, C.G.
[2]*Human Resource Development Centre,*
Pt. Ravishanker Shukla University, Raipur, C.G.

ABSTRACT

The purpose of present work is the identification and documentation of ethno medicinal weeds growing on rice fields of Rajim region. Our survey was aimed at the possibilities of discovering new ways by which such plants could be better utilized for the welfare of human health. Weeds are generally defined as unwanted or undesirable and easily grown vegetation. But weeds are not really unwanted especially in terms of traditional herbal medicines. In the written record, the study of herbs dates back over 5000 year to the Sumarian who described well established medicinal uses for plants. Several workers studied the medicinal properties of plants viz Gangwar et.al.(2010), Pati and Agrawal (2010), Sinha (2013) to understand the uses of plant species to cure various ailments of mankind. According to Gangwar et.al villagers and tribes still use medicinal herbs for treatment of common ailments like cold, cough, fever, headache, body ache, constipation, dysentery etc. The present study has been done by random sampling method. A total 64 species of plants representing 56 genera and 21 families were collected. Some dominant weeds on rice fields are Alternanthera sessilis L., Achyranthus aspera Linn, Blumea tacera (Burn.f.), Cynodon dactylon pers, Cyperus rotundus L Eclipta prostrate.L. urban, Euphorbia hirta L., Phyllanthus niruni Webster, Sida cardifolia. L. Sphaeranthus indicus. L. Tridex procumbance (linn), Tephrosia purpurea. L. (pers),. It was observed that leaves, roots, stem bark are used for various ailments whole part of the plant is used maximum for treatment. Traditional uses were obtained from inhabitants of the area,ayurvedic doctors and a study of the pertinent literature. However the collections of these medicinal weeds provide farmers with a most welcomed additional income.

Keywords*: Ethnomedicinal weeds, Rice fields, Rajim region, Chhattisgarh.*

Introduction

Rice is a staple food for more than 60 per cent of the world's, seven billion people and more than 90 per cent of the rice is consumed in Asia (Mohanty 2013, Chauhan *et al.*, 2014).

Chhattisgarh is agricultural chief land and due to large production of rice Chhattisgarh is commonly known as Rice Bowl of India (Dhan ka Katora). Chhattisgarh has been divided into three agro – climatic regions (i) Plain Chhattisgarh, (ii) Plateau of Bastar, (iii) Northern Hilly region. In Plain Chhattisgarh region 81 per cent land is under rice cultivation, while the other two plateau of Baster and Northern region – have 71 per cent,64 per cent land respectively under rice. The weed flora in rice field is very much diverse and dynamic over times and place. So what is weed? A Plant that is growing where it is not wanted (Roberts *et al.*, 1982), A plant out of place (Bletchley 1912), A plant that is glowing where it is desired that something else growth (Georgia 1916), These naturally glowing plants are generally referred as weeds. But weeds are not really unwanted especially in terms of traditional herbal medicines. In earlier times, all drugs and medicinal agents were derived from natural substance. Most of these remedies were obtained from wild herbs and used against various ailments like cold, cough, fever, headache, constipation, dysentery etc. Several workers studied the medicinal properties of plants viz Arvind *et al.* (2005), Gangwar *et al.* (2010), Pati and Agrawal (2010), Sinha (2013), to understand the uses of plant species to cure various ailments.

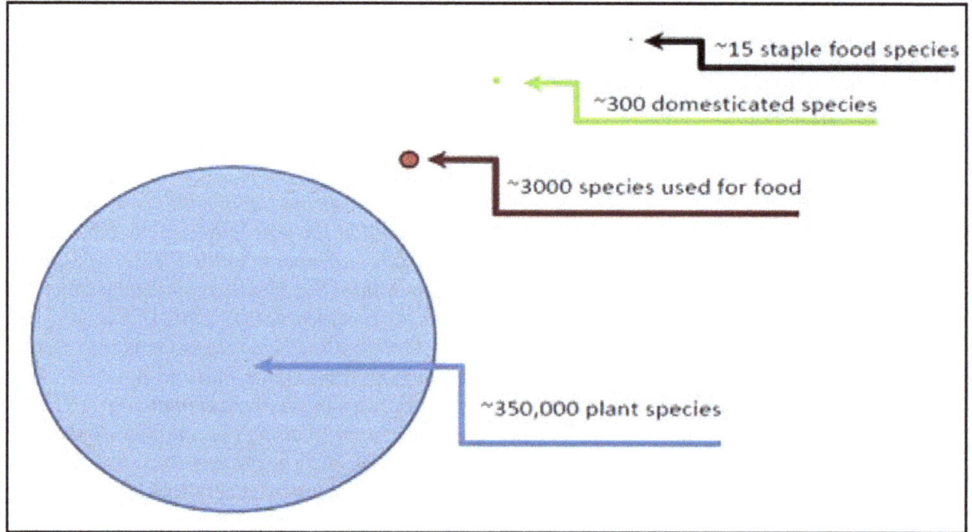

Figure 7.1: How many Weed Species are threre?

Materials and Methods

The present study was carried out in Rajim Region *viz.* located at 20.9632° N and 81.8907°E latitude, Elevation – 281m.

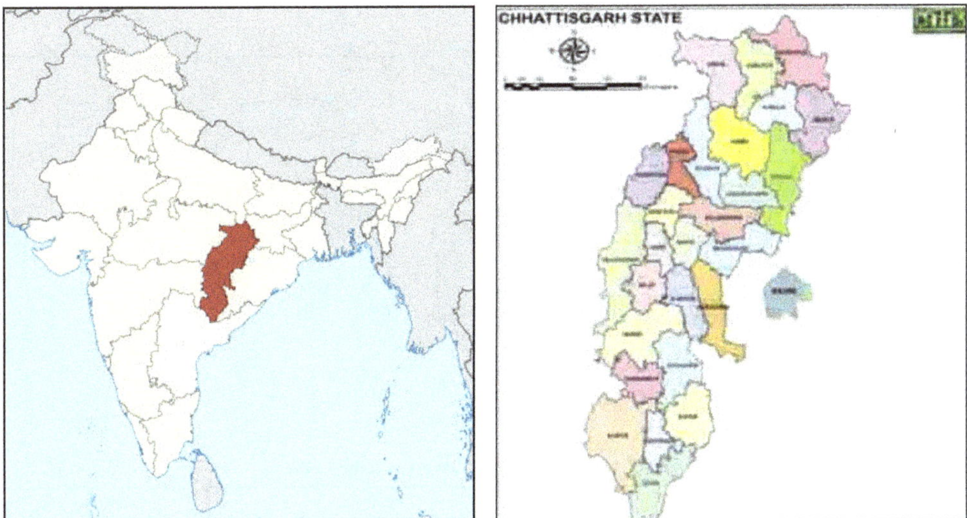

Figure 7.2: Study Area.

Average Highest Temperature of Rajim is 42° C in month of May and Average Lowest Temperature is 13 °C in month of January.

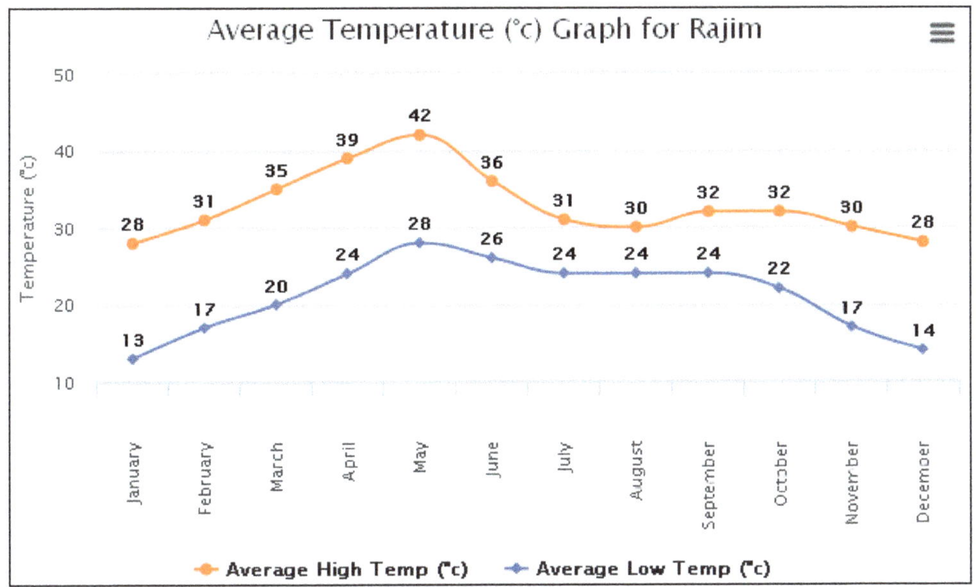

Figure 7.3: Average High/Low Temperature for Rajim, India.

Average Hight Rainfall 310 m.m. is in average 24 days and Average Lowest Rainfall 0.5m.m. in 2 days.

Present study has been done by random sampling method. Only weedy herbs of unit area had been study. Collected plants were broad to laboratory for botanical

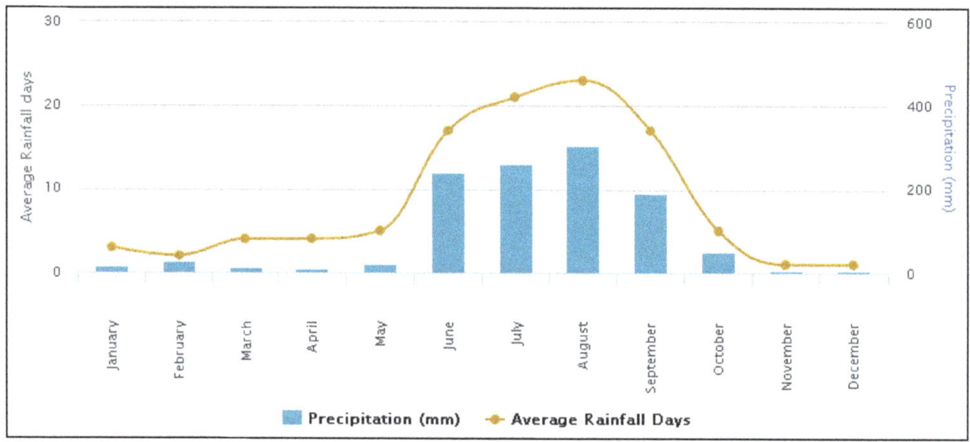

Figure 7.4: Average Rainfall (mm Graph for Rajim).

diagnosis, their detailed information's pertaining to the botanical name, family name, and medicinal use etc. Collected plant species were maintained in the form of herbarium.

Observation and Result

In present study total 64 plant species were recorded. These plants having curative properties these are showed in Table 7.1. It was observed that roots, stem, leaves, flowers and seeds are used for various ailments. The list of Weed Plants found in Rice Fields of Rajim Region is shown in Table 7.1.

Table 7.1: List of Weed Plants

Sl.No.	Botanical Name of the Plant	Family	Plant Parts Used	Treatment
1.	*Abutilon indicum* (L.)	Malvaceae	Leaves	Cold, Cough, Indigestion
2.	*Achyranthes aspera* L.	Amaranthaceae	Whole part of plant root, seed, leaf	Dissolve stone, piles, abdominal colic pain, fibroids, glandular growths.
3.	*Aerva lanata* L. Jussex schutt.	Amaranthaceae	Whole plant	Snake bites, kidney stones, Alzheimer, Anemia
4.	*Alternanthera paronychoides* A. St. -Hil	Amaranthaceae	Whole plant	Antioxidant, antiapoptotic
5.	*Alternanthera Sessilis* (L.)	Amaranthaceae	Whole part of plant	Eye complication
6.	*Alysicarpus monitifer* (L.) DC.	Fabaceae	Whole plant	Liver disorder, diabetes, an therosclerosis, inflammatory disease
7.	*Amaranthus spinusus* L.	Amaranthaceae	Whole plant	internal bleeding, diarrhea, excessive menstruration, snake bite, boils, vaginal discharge.
8.	*Amaranthus viridis* L.	Amaranthaceae	Whole plant	Luxative, stomachic, appetizer, anti pyretic, leprosy, bronchitis, piles, leucorrhoea, constipation
9.	*Ageratum conyzoides* L.	Asteraceae	Whole plant, roots	Analgesic effect, colic, cold, fever, insecticidal activity
10.	*Argemone mexicana* l.	Papavaraceae	Leaves, stem, root	Diuretic, worms, cut and healing
11.	*Asteracantha longifolia (L.)*	Acanthaceae	Whole plant	Cold, cough, jaundice, anemia, baldder stone, kidney stone
12.	*Borraavia diffusa* L.	Nyctaginaceae	Whole part of plant	Insect bite, cuts, wounds and fever
13.	*Barlaria primonitis* L.	Acanthaceae	Leaves	Leucoderma
14.	*Blumea tacera* (Burn.f.)	Asteraceae	Leaves and whole plant	diuretic, anti scorbutic, febrifuge, anthelmintic
15.	*Calotropis gigantea* (L.)	Asclepiadaceae		Fevers, rhematism, indigestion, cough, cold, eczema, asthma, elephantiasis, nausea, vomiting
16.	*Calotropis procera* ait. R.Br.	Asclepiadaceae	Whole part of plant	Cough, cold, echzema, epilepsy, antioxidant
17.	*Cassia tora* L.	Caesalpiniaceae	Leaf, root and seeds	Eczema, dematomycosis, dandruff, constipation cough, liver disorder, blood purifier, nervetonic
18.	*Convolvulus piuricaulis* L.	Convolulaceae	Whole part of plant	Pain and inflmation

Contd...

Table 7.1–*Contd...*

Sl.No.	*Botanical Name of the Plant*	*Family*	*Plant Parts Used*	*Treatment*
19.	*Cynodon dactylon* Pers.	Poaceae	Leaves	Vomiting, leprosy, scabies, skin diseases, fever and dysentery
20.	*Cyperus rotundus*	Cyperaceae	Rhizorne, roots tubers	Rhizome are considered diuretic, analgesic, carminative, sedative, stoachic, vermifuge and used in skin disease.
21.	*Caesulia axillaris* Roxb.	Asteraceae	Whole plant	Antialfatoxigenic, cold, cough, nasal congestion, malaria, antifungal and anti aflatoxigenic.
22.	*Celosia argentea* L.	Amaranthaceae	Whole plant	Dysentery, excessive menstruatum, intestinal bleeding, retinal hemorrhage, conjunctivitis, eye diseases
23.	*Commelina benghalensis* L.	Commelinaceae	Whole Plant	Skin diseases, fever, wounds, boils and vitality
24.	*Croton bonplandianum* Baill.	Euphorbiaceae	Stem leaves, seeds, root	Antiseptic, treatment of cholera
25.	*Cynotis axillaris* (Linn.) Schult.	Commelinaceae	Whole plant	Tympanitis, external application in ascites and boils, rheumatisms and joint pains.
26.	*Cynotis cristata* (L.) D. Don.	Commelinaceae	Roots	Swelling and snakebite
27.	*Dactyloctenium aegyptium* (L.) Beauv.	Poaceae	Leaves and whole plant	Decotion of whole plant used in lumbago, accelerate child birth, dysentery, cure of ulcers.
28.	*Datura metal* L.	Solanaceae	Seeds	Asthma, Skin problem, narcotic, sedative
29.	*Desmodium triflorum* (L.) D.C.	Fabaceae	Root	Dysentery, stone removal
30.	*Echinochloa colona* (L.) Link	Poaceae	Seeds	Cures Ingestion
31.	*Eleusine indica* (L.) Gaertn	Poaceae	Whole plant, root	Anthelmintic, astringent, depurative, diuretic, febrifuge, bladder disorder, liver complaints
32.	*Eclipta alba* L.	Asteraceae	Whole plant	Cold, cough, echzema, epilepsy.
33.	*Euphorbia hirta* L.	Euphorbiaceae	Whole plant	Female disorders, respiratory ailments (cough, bronchitts and asthma) dysentery, jaundice.
34.	*Euphorbia hypercifolia*	Euphorbiaceae	Whole plant	Treatment of sexually transmitted infections, asthma bronchitis.

Contd...

Table 7.1–*Contd...*

Sl.No.	Botanical Name of the Plant	Family	Plant Parts Used	Treatment
35.	*Euphorbia cyathophora*	Euphorbiaceae	Whole plant	Antimicrobial and wound healing activities
36.	*Evolvulus alsinoides* L.	Convolvulaceae	Whole plant	Alzheimer, anxiety, blood impurity, antidepressant, antioxidant
37.	*Echinops echinatus* Roxb.	Asteraceae	Root leaves fruits and bark	Stimulant to treat sexual debility, antitungal, analgesic, diuretic, hepatoprotective, antiinflammatory, wound-healing, anti pyretic
38.	*Fimbristylis aestivalis*	Cyperaceae	Whole plant	Used as poultice on inflammations.
39.	*Fimbristylis milliacea* (Linn)	Cyperaceae	Roots and leaves	Dysentery, poulticing in fever
40.	*Gisekia pharnaceoides* L.	Gisekiaceae	Whole plant	Warts, asthma, after miscarriage, antibacterial, C.N.S. depressant and antithelmintic activity
41.	*Hyptis suaveolens* (L.) Poit	Lamiaceae	Whole part of plant, root, leaves	Protects from gastric ulcer,fever, cold, antifungal, boils, eczema.
42.	*Indigofera caerulea* Roxb.	Fabaceae	Leaves and roots	Wound dressing, constipatim, infected eyes.
43.	*Lantana camara* (Linn)	Verbenaeeae	Whole part of plant	Bronchitis antifungal, antiulcerogenic, hemolytic anti-hyperglycemic, antioxidant, antiurolithiatic activity
44.	*Leucas aspera* L.	Lamiaceae	Whole plant	Stop nauseating feeling, antimicrobial, anti fungal activities, anti pyretic, to treat scorpion bites.
45.	*Marsilea quadrifolia* L.	Marsileaceae	Leaves, whole plant	Antiinflammatory,diuretic,depurative,febrifuge,refrigerant.
46.	*Mimosa pudica* L.	Fabaceae	Whole plant	Uicer,piles,diarrhea,antivenom activity.
47.	*Merremia emarginata* (Burm.f)	Malvaceae		Diuretic, anthelmintic, carminative, cardiac disease, gastropathy, nephropathy, rat. bite, leucoderma
48.	*Oxalis corniculata* L.	Oxalidaceae	Whole part of plant	Piles, whole plant is good appetizer
49.	*Parthenium hysterophorus* L.	Asteraceae	Leaves, root	Skin problems, allergy
50.	*Phyllanthus niruri* Hook f.	Euphorbiaceae	whole part of plant	Jaundice
51.	*Portulaca olerocea* L.	Portulacaceae	Leaves, whole plant,	Antioxidant, abnormal uterine bleeding, asthma, type-2 diabetes

Contd...

Table 7.1–*Contd...*

Sl.No.	*Botanical Name of the Plant*	*Family*	*Plant Parts Used*	*Treatment*
52.	*Rungia pectinata* (L.) Ness	Acanthaceae	Leaves, whole plant, root	Antinfilammatery, diuretic, anti microbial small pox, cut, wounds
53.	*Sida acuta* Burm.f.	Malvaceae	Leaves	Dibetes, dysentery, nerve weakness
54.	*Sida quardifolia* L.	Malvaceae	Leaves	Nerve weakness
55.	*Solanum nigrum* L.	Solanaceae	Whole part of plant	Blood purifier, treatment of ulcer chronic fever, arthritis, antipyretic, jaundice
56.	*Solanum xanthocarpum* Schrad. and H. Wendl	Solonaceae	Root, leaves, fruits whole plants	Bronchial asthma, digestive disorder, piles, anti cancer and anti HIV perspective
57.	*Sonchus oleraceus* L.		Whole plant	Diarrhea, menstrual problems, fever, warts inflammation.
58.	*Sphaeranthus indicus* L.	Asteraceae	Leaves, flower	Toothache, epilepsy mental illness, hepatopathy, jaundice, diabetes, leprosy.
59.	*Spilanthus acmella* F. Linn	Asteraceae	Leaves, flower	Toothaches, stammering stomatitis, diuretic action in rats.
60.	*Saccharum spontaneum* L.	Poaceae	root	Astringent, emollient laxative, aphrodisiac and tonic
61.	*Tephrosia purpurea* (Linn.) Pers	Fabaceae	Whole plant	Leprosy, ulcer, asthma, tumors, diseases of liver, spleen, heart and blood
62.	*Tridax procumbens* L.	Asteraceae	Whole part of plant	Stop wound bleeding, free motion, headache
63.	*Vernonia cinerea* L. Less	Asteraceae	Leaves	Fileria, leucoderma
64.	*Xanthium strumarium* L.	Asteraceae	Whole part of plant	Toothaehe, headache, malaria, fever

Rungia pectinata **(L.) Ness** ***Tephrosia purpurea*** **(Linn.) Pers.**

Spilanthus acmella **F. Linn.** ***Asteracantha longifolia*** **(L.)**

Figure 7.5: Some Weeds.

Discussion and Conclusion

During the exploration 64 members of plant species under 56 genera belonging 21 families along with one pteridophyte. Out of 64 Members of plant species 56 species of dicots are distributed under 49 genera belonging to 19 families, while 08 species of monocot.10 Mambers of Asteraceae, 07 Mambers of Amaranthaceae, 05 Mambers of Poaceae and Euphorbiaceae, 04 Mambers of Malvaceae were found. Out of 21 families 08 families are represented by single species. Asteraceae, Amaranthaceae, Euphorbiaceae, Poaceae and malvaceae are the largest family respectively.

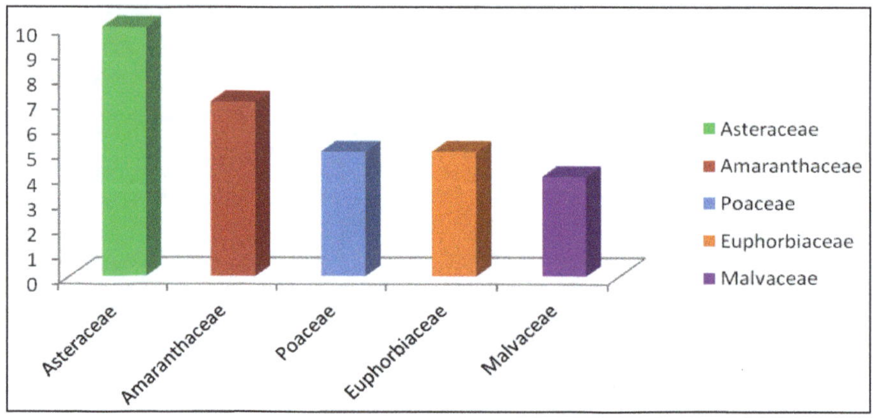

Figure 7.6: Five Dominant Families.

Percentage of families, genera and species of dicots and monocots are represented below:

Groups	*Families*		*Genera*		*Species*	
	No.	*Per cent*	*No.*	*Per cent*	*No.*	*Per cent*
Dicotyledones	19	90.47	49	87.5	56	87.5
Monocotyledones	02	9.5	07	12.5	08	12.5

Herbal medicines possessing bio-active chemicals without any side effect and hence they are very much popular. *Achyranthes aspera, Calotropis sps., Tridax procumbens, Lantana camara, Cynadon dactylon are the leading species* used as remedies against various ailments. Most of the species are used as antirheumatic, antipyretic, for respiratory disorders, skin diseases, hemorrhoids and as diuretic. This attempt describes traditional uses of the medicinal weeds and additional source of income for formers

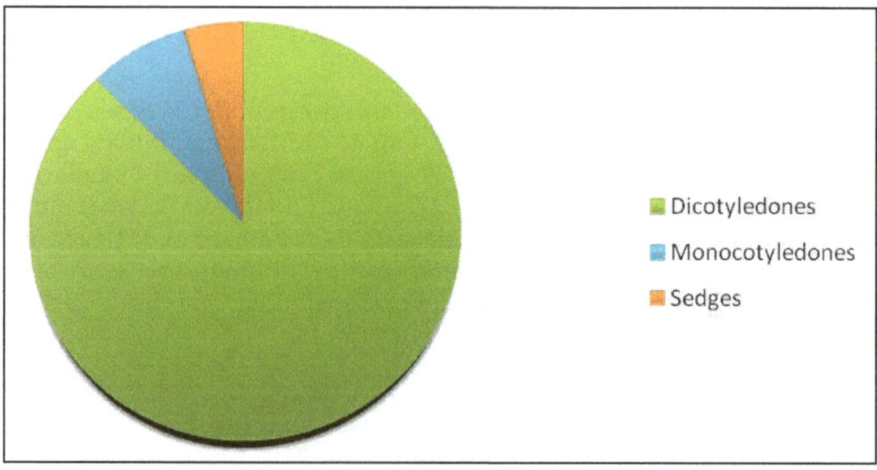

Figure 7.7: Distribution of Weed Species.

References

Agrawal S. C. and Pati R. N. (2010). Folk medicine folk weather and medicinal plants of Chhattisgarh. Sarup. Book Publication Pvt. Ltd.

Arvind *et al.* (2005). Are Red listed spp. Treatened; A comparative analysis of red listed and non-red listed plant species in the western ghat, *India current science* Jan 2005.Vol. 88; No- 2 25.

Baranwal Akanksha, Mazumdar Arijit, Chakraborthy G.S., Gupta Seema 2014, Phytopharmacological uses of *Tephrosia pupurea – A review, pharmocophore,* vol 5(4), 658-665.

Chan K, Islam MW, Kamil M., Radhakri Shran R, Zakaria M.N., Habibullah M, Atlas A.2000. The analgesic and anti inflammatory effect of *portulaca oleracea L.* Sub Sp. Sative (Haw.) Celak, *J Ethnopharmacol,* 2000 Dec; 73(3) 445-51.

Chitra, M. N Senthikumar, M Asraf Ali 2014 Anti microbial and wound healing activities of *Euphorbia cyathophora. International Journal of Pharmacology Research.* 2014 Vol 4 (1) 59-61.

Chopra R.N., Nayer S.L. and Chopra I.C., Glossary of Indian Medicinal Plants (Including the supplement) Council of Scientific and Industrial Research, New Delhi 1986.

Filipps, De R.A.; Maina, S.L., Crepin, *J. Medicinal plants of the guianas.*

Galani, V.J. Patel, B.G. and Rana, D.G. 2010. *Sphaeranthus indicus* Linn: A phytopharmacological review. *International Journal of Ayurveda Research,* 1(4), 247-253.

Gangwar, *et al.,* 2010. "Ethenomedicinal plant diversity in kuman, Himalaya of Uttrakhand India". *Natures and Sci.* 2010. 8 (5).

Hamiltone, 1852, The flora Homeopathica, Reprint Edition (1997). Pbl. B Jain Publisher Pvt. Ltd., New Delhi, pp 147-149.

Kalita, Sanjeeb Gaurav Kumar, Karthik Loganathan Bhaskar Rao, 2012 A Review on Medicinal Properties of *Lantana camara Linn. Research Journal of Pharmacy and Technology,* July 2012.

Karthikeyan, K. C.K. Dhapal and Gopala Krishnan. 2015 Preliminary phylochemical investigation of whole plants of *Alysicarpus monilifer* (L.) DC. *International Journal of Pharma and Biosciences* 2015 Jan 6(2): pp 67-72.

Khair, Abul Mohammed Ibrahim, Qamrul Ashan, Mohammad A. Rashid, Pharmacological Activities of *Blumea lacera* (Burn.f) DC: A medicinal plant of Bangladesh. 2014 July, SCIEN 4(13).

Lim Y.Y. and Quah E.P.L. 2007. Antioxidant properties of different cultivars of *portulaca oleracea food chemistry* 103 (3):734-740.

Maurya, Santosh Kumar Ashwini Kumar Kushwaha and Ankit Seth, 2015, Ethnomedicinal review of Usnakantaka (*Echinops echinatus Roxb.*), *Pharmaco gnosy Rev* Jul-Dec; 9(18) 149-154.

Ming L.C. (1999). *Ageratum Conyzoides*: A tropical sources of medicinal and agricultureal products p.469-473, In: J. Janick (ed), perspectives on new crops and new uses. ASHS Press, Alexandria, VA.

Panigrahi Ashok. K. 2000. "Glossary of useful and economically important plant". New central book agency (P.) Ltd.

Prajapati, M.S. J.B. Patel, K. Modi and M.B. Shah, 2010 *Leucas aspera: A review pharmacogn Rev.* 2010 Jan-Jun; 4(7):85-87.

Puratchikody A, Devi C.N. and Nagalakshmi G. 2006. wound healing activity of *Cyprus rotundus* Linn *Ind. J. Pharm. Sci* 68:97-101.

Purushoth Prabhu T., Panneer Selvam M.P, Selvakumar S., Udhumansha Ubaidulla, Shantha.2012 A. Anticancer Activity of *merremia emarginata* against Human Cervical Breast Carcinorna. *International Journal of Research and Development in Pharmacy and Life science* oct-nov; 2012, Vol 1, No. 4, pp 189-192.

Sardar N.Y.M., Sardar N.U., Sanzida M. and Muhammad A.A. 2008, Antioxidant and Antibacterial Activities of *Calotropis procera* Linn. *American – Eurasion J. Agric Envirm Sci* 4(5): 550-553.

Sathya, M. R.K. Kokilavani, 2012 Antilithiatic Activity of *Saccharum spontanuem* Linn on Ethylene glycol- Induced Lithiasis in Rats. *International Journal of Pharma Science and Research* 2012 sep. Vol 3 No. 9 pp 467-472.

Shashank Kumar and Abhay K. Pandey 2014. Medicinal attributes of solanum xanthocarpum fruit consumed by several tribal communities as food: an in vitro antioxidant, anticancer and anti HIV perspective, *B.M.C. Complementary and Alternative* Medicine 14:112.

Simpson D.A., Inglis C.A. Cyperaceae of Economic Ethmbotanical and Horti Cultural Importance: A Checklist Kew Bulletin Vol. 56 No. 2, pp 257-360.

Sing, U Wadhwani, AM and Johri, B.M. (1996). Dictionary of Economic plants in India. Pbl Indian Council of Agricultural Research, New Delhi, p. 165.

Sinha M. K. 2013. "Treat assessment of medicinal plants of Korea district in Chhattisgarh" (India).

Swain, S.R. B.N.Sinha, and P.N. Murthy, 2008. Anti-inflammatory, Diuretic and antimicrobial activities of *rungia pectinata* Linn and *Rungia repens* Ness. Indian *Journal of Pharmaceutical Sciences*. 2008 Sep. 1-18.

Swapna M.M., Prakash Kumar R., Anoop K.P., Manju C.N. and Rajith N.P.- 2011. A review on the medicinal and edible aspects of aquatic and wet lond plants of India. Journal of medicinal plants research 2011 Dec, Vol. 5 (33) pp. 7163-7176.

U. Yasmin Stella; E. Sasikala, Dr. G. Srinivasa Rao and J. Sangeetha. 2004 Pharmacognostic studies on *Gisekia pharnace Oldes Linn, Anc Sci Life*. Jan-Mar; 23(3): 13-21.

Uphof J.C., 1959. Dictionary of economic plant P. 267 (Englamann Weintherim). Wet, H. De Y.N. Nzaona, S.F. Van Vuuren. 2012 Medicinal plants used for the treatment of Zprovince, South Africa, South African. *Journal of Botany* 2012 78: 12-20.

8

Study of the Plants of Gandhi Udyan of Raipur City

R. Verma[1], P. Dewangan[2], V. Acharya[3] and A.K. Girolkar[4]

[1]Student, M.Sc. IVth Sem (Botany) 2014-15
[2]Research Scholar,
[3]Assistant Professor, Botany,
[4]Principal,
Govt. D.B Girls P.G. College, Raipur, C.G.

Introduction

Chhattisgarh state was formed on 1 November 2000 by partitioning 16 Chhattisgarhi speaking southeaster districts of Madhya Pradesh. Raipur was made its Capital city. Raipur is located near the centre of a large plain, sometimes referred as the "Rice bowl of India", where hundreds of varieties of rice are grown. Chhattisgarh has 3 National Parks and 11 wild life sanctuaries. Chhattisgarh have 2nd largest forest cover *i.e.* 44 per cent in India.

Raipur is located near the centre of a large plain. Raipur has a tropical west and dry climate, temperatures remain moderate throughout the year except from march to June, which can be extremely hot. The temperature in April-May sometimes rises upto 48°C (118°F). The city receives about 1,300 millimeters (51 inch) of rainfall, mostly in the monsoon season from late June to early October. Winters last from November to January and are mild, although lows can fall to 5°C (41°F).

Study Area

The studied area is Gandhi Udyan which is managed by Raipur Municipal Corporation (RMC). Udyan is located in the heart of city near Bhagat Singh chowk.

Udyan is spread over an area of 4.5 acres. Udyan was renovated by plantation of trees, installation of playground equipment for children and lighting displays.

Phytodiversity concerns with the variety and variability of plant species. The studied area consists of perennial trees, cultivated shrubs, ornamental plants as well as different types of weeds. As mentioned above area represents floral diversity. It includes ornamental, medicinal and economically important plants.

Materials and Methods

Survey in the study area has been made between January 2014 to October 2015. Plants were identified with the help of flora of Hooker 1872-97 Oommachan, M. (1976)., Tiwari K.P. (1995),Verma, D.M., N.P. Balkrishan and R.D. Dixit (1994. Also habit and habitat of plants were recorded. Pictures were taken and plant samples were collected.

Table 8.1

S.No.	*Vernacular Name*	*Botanical Name*	*Family*	*Habit*	*Wild Cultivated*
1.	Babool	*Acacia melanoxylon* R. Br.	Leguminosae	Trees	Wild
2.	Senegal	*Acalypha ciliate* Forsk.	Euphorbiaceae	Herb	C
3.	Indian nettle	*Acalypha indica* L.	Euphorbiaceae	Herb	W
4.	Bel	*Aegle marmelos* (L.) Corr.	Rutaceae	Tree	C
5.	Century plant	*Agave americana* L.	Asparagaceae	Shrub	C
6.	Goat weed	*Ageratum conyzoides* L.	Asteraceae	Herb	W
7.	Siris	*Albizia lebbeck* (L.) Benth.	Leguminosae	Tree	C
8.	Mahaneem	*Alianthus excelasa* Roxb.	Simaroubaceae	Tree	W/c
9.	Chattim	*Alstonia scholaris* (L) R. Br.	Apocynaceae	Tree	C
10.	Neem	*Azadirachta indica* Adr. Juss.	Meliaceae	Tree	W/C
11.	Bamboo	*Bambusa polymorpha*	Poaceae	Tree	C
12.	Kesariya	*Barleria prionitis* L.	Acanthaceae	Herb	C
13.	Kachnar	*Bauhinia purpurea* L.	Leguminosae	Tree	C
14.	Sindoor	*Bixa orellana* L.	Bixaceae	Shrub	W/C
15.	Semal	*Bomabax malabaricum* DC	Bombaceae	Tree	W
16.	Kapok	*Bombax ceiba* L.	Bombaceae	Tree	W/C
17.	Kagaj phool	*Bougainvillea glabra* choisy.	Nyctaginaceae	Shrub	C
18.	Palash	*Butea monosperma* (Lam.) Taubert	Leguminosae	Tree	W
19.	Weeping Bottlebrush	*Callistemon viminalis* (Sol. ex Gaertn.) G.Don	Myrtaceae	Tree	C
20.	Madar	*Calotropis gigantea* (L) R.Br.	Asclepiadaceae	Tree	W
21.	Vaijanti	*Canna indica* L.	Cannaceae	Herb	C
22.	Papita	*Carica papaya* L.	Caricaceae	Tree	C

Contd...

Table 8.1–*Contd...*

S.No.	Vernacular Name	Botanical Name	Family	Habit	Wild Cultivated
23.	Sulfi	*Caryota urens* L.	Palmae	Tree	W/C
24.	Amaltas	*Cassia fistala* L.	Leguminosae	Tree	//C
25.	Sadabahar	*Catharanthus roseus* (L.) Don	Apocynaceae	Herb	C
26.	Coconut	*Cocos nucifera* Linn.	Palmae	Tree	C
27.	Cycas	*Cycas circinalis*	Cycadaceae (Gymnosperm)	Tree	C
28.	Sago palm	*Cycas revoluta* Thunb.	Cycadaceae	Tree	C
29.	Doob grass	*Cynodon dactylon* (L.)Pers.	Poaceae	Grass	W
30.	Shisham	*Dalbergia sissoo* Roxb	Leguminosae	Tree	C
31.	Gulmohar	*Delonix regia* (Hook)raf	Leguminosae	Tree	C
32.	Golden dew drop	*Duranta repens* L.	Verbenaceae	Herb	C
33.	Amla	*Emblica officinalis* Gae.	Euphorbiaceae	Tree	C/W
34.	Nilgiri	*Eucalyptus lanceolatus*	Myrtaceae	Tree	C
35.	Mottled spurge	*Euphorbia lactea* Haw.	Euphorbiaceae	Tree	C
36.	Bargad	*Ficus benghalensis* L.	Moraceae	Tree	W
37.	Peepal	*Ficus religiosa* L.	Moraceae	Tree	W
38.	Bottle palm	*Hyophorbe lagenicaulis* (L.H.Bailey) H.E. Moore	Palmae	Tree	C
39.	Subabool	*Leucaena leucocephala* (Lam.) de Wit	Leguminosae	Tree	W
40.	Chinese fan palm	*Livistona chinensis* R Brown	Palmae	Tree	C
41.	Aam	*Mangifera indica* L.	Anacardiaceae	Tree	C
42.	Maulshri	*Mimusops elengi* L.	Sapotaceae	Tree	W/C
43.	Red Kaner	*Nerium odorum* Aiton.	Apocynaceae	Shrub	C
44.	Parijat	*Nyctanthes arbor-tristis* L.	Oleaceae	Tree	C
45.	Tulsi	*Ocimum sanctum* L.	Lamiaceae	Herb	W/C
46.	Wood Sorrels	*Oxalis corniculata.* L.	Oxalidaceae	Herb	W
47.	Copperpod	*Peltophorum ferrugineum* Benth.	Leguminosae	Tree	W/C
48.	Bhui amla	*Phyllanthus niruri* L.	Phyllanthaceae	Herb	W
49.	Gangalmli	*Pithecellobium dulci* (Roxb)Benth.	Leguminosae	Tree	W
50.	False ashok	*Polyalthia longifolia* (Sonn) Thwaites	Annonaceae	Tree	C
51.	Karanj	*Pongamia glabra* Vent.	Leguminosae	Tree	W
52.	Almond	*Prunus dulcis* (Mill.) D. A. Webb	Rosaceae	Tree	W/C
53.	Rugtoora	*Spathodea campanulata* P. Beauv	Bignoniaceae	Tree	C/W

Contd...

Table 8.1–*Contd...*

S.No.	Vernacular Name	Botanical Name	Family	Habit	Wild Cultivated
54.		*Sporobolus diandrus* (Retz.) P.Beauv.	Graminae	Grass	W
55.	Wild almond	*Sterculia foetida* L.	Sterculiaceae	Tree	W
56.	Jamun	*Syzygium jambolanum* DC	Myrtaceae	Tree	W/C
57.	Imli	*Tamarindus indica* L.	Leguminosae	Tree	W
58.	Yellow elder	*Tecoma stans* (L.) Kunth.	Bignoniaceae	Tree	C
59.	Sagon	*Tectona grandis* L.	Verbenaceae	Tree	/C
60.	Yellow kaner	*Thevetia nerifolia* juss.	Apocynaceae	Shrub	C
61.	Vidya	*Thuja occidentalis* L.	Cupressaceae	Tree	C
62.	Bengal clockvine	*Thunbergia grandiflora* (Roxb. ex Rottler) Roxb.	Acanthaceae	Vine	C
63.	Coatbuttons	*Tridax procumbens* L.	Asteraceae	Herb	W
64.	Ironweed	*Vernonia* sps. Schreb	Asteraceae	Herb	W
65.	Ber	*Ziziphus jujuba* Mill.	Rhamnaceae	Tree	W

Observation and Results

During the survey about 65 plant species belonging to different families have been identified. Of which some herbs, some shrubs, climbers, xerophytic plant and trees are recorded. Out of 65 plant species 12 herbs, 5 shrubs, 1 climber and 45 tree species are identified. 2 grasses and 2 Gymnosperm are also recorded. Despite being presence of ornamental plants udyan also shows presence of different grasses and exotic weeds.

Azadirachta indica

Butea monosperma

Carica papaya

Caryota urense

Cassia fistula

Cycas revoluta

Emblica officnalis

Euphorbia lactea

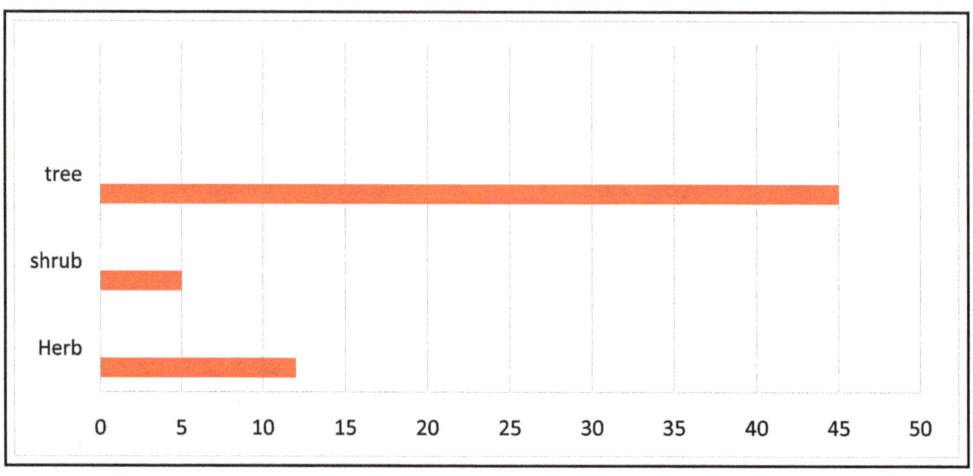

Conclusion

People visit udyan or park for recreation, morning walk, jogging and exercising. Some plants of the udyan are ethnobotanically important plants. Some plants are economically important ones and non-wood forest producers. Most of the plants are soil binders. Trees helps maintain low temperature in the park and work as air conditioning system. Vegetation in the park give a sense the aesthetic sense to the visitors. The park posses exotic, indigenous, wild and cultivated plant species. *Caryota urense* is famous as "Bastar Beer " and due to its showy inflorescence it is also planted as ornamental plant. *Euphorbia lactae* is small xerophytic tree with showy appearance. This park represents an example of *ex-situ conservation* of many plant species.

References

Hooker, J.D. (1872-97). The flora of British India, Vols I – VII Reeve and Co., London, England.

Nair, N.C., and K.K. Khanna (2005). Floristic Diversity of Chhattisgarh. Biden Singh Mahendra Pal Simgh, Dehradun.

Oommachan, M. (1976). Flora of Bhopal, J.K. Jain Brothers, Bhopal.

Tiwari K.P. (1995). Flora of Amarkantak for detailed project formation to constitute Amarkantak Biosphere Reserves. Draft Project Report, SFRI, Jabalpur.

Verma, D.M., N.P. Balkrishan and R.D. Dixit (1994). Flora of Madhya Pradesh Vol. I, Botanical Survey of India. Calcutta. p.p. 1-668.

Diversity of Medicinal Plants in Chhindwara District

Simpal Patil

Department of Botany,
Rajmata Sindhia Govt. Girls College, Chhindwara, M.P.

ABSTRACT

The Chhindwara district of (M.P.) has immersed forest area covered by numerous plants having medicinal properties. It reflects richness of floral diversity in term of their ethno botanical importance. During the year form 2013 to 2015 an extensive field work on medicinal plant has been carried out in different localities of Chhindwara district. Medicinal plants have been an integral part of health and healthcare for centuries and will continue to play a vital role in public health for generation to come. As per WHO, about 80 per cent population in developing countries relies on traditional natural medicines and almost 80 per cent of the traditional medicines involve the use of plant extracts. According to a recent survey, herbal preparations have been found to be more popular in primary healthcare in China, Malaysia, Nepal, Myanmar, Belgium, France, Germany and Netherland. The developed countries are also turning towards the use of traditional medicinal systems that involve the use of herbal drugs and remedies. Thus, the continuously growing demand of the phytomedicines in the International market has been exerting and outstripping pressure on the Indian phytoresources. In Chhindwara district more than 250 species of medicinal plants distributed throughout the district in which numerous species are traditionally and ethno medicinally used by the local tribal. They may be lost if their medicinal aspects are not properly documented.

Introduction

Biodiversity of traditional medicinal herbs and plants is continuously under the threat of extinction as a result of growth exploitation, environment unfriendly harvesting techniques, loss of growth habitats and unmonitored trade of medicinal plants. In other word, biodiversity resources such as medicinal and other plants are the foundation for progress is strengthening food, health and livelihood security. The diversity of these resources also act as feedback for modern biotechnology and industry. Effectively used, it can become a major instrument for economic prosperity of the state as well as communities.

The history of medicine dates back probably to the origin of human race on this earth. In India, the reference to medicinal properties of some herbs in 'Rig-Veda' seems to be the earliest record of the use of plants in medicine. But there references are very brief. More detailed account is available in 'Atharva Veda'. After this period, these seem to be a gap of about thousand years in this direction. Later on two most valuable works on Indian system of medicine *viz.* 'Charak Samhita' and Sushruta Samhita' appeared, which dealt with about 700 drug plants.

World Health Organization has estimated that by the turn of this century, about 80 per cent of the world's population will rely on plant based medicines for their healthcare. While the importance of medicinal plants and related research is growing their habitats are shrinking and their numbers declining. In India, this decline has been rapid and has resulted in an acute shortage of indigenous medicines and the substitution of genuine plants by spurious alternatives.

Study Area

The study site comprise of various forest communities spread all over Chhindwara (M.P.) district. The district is located in the southern Madhya Pradesh encompassing latitude 20 10' to 22 30' North and longitude 78 0-58 0 longitude at 685 above mean sea level in the north Betul district in the West Seoni district and the southern part is a boundary of Maharashtra state.

The total geography area is 8332 sq.km. and 48 per cent of total area of the earth is surrounded by forest.

Source of Data

The district is presently divided into three forest divisions namely east forest division. West forest division and north forest division. The tehsil of district is Parasia, junardeso, Chourai, Harrai, Tamia etc. The forest is of high quality with mixed miscellaneous nature, with more or less contiguous and compact patches. However, observations reveal that the Reserve Forest category of North except in the western part. Most of the protected forest category is in close proximity of the reserve forest category.

The district has vast forest resources and mineral wealth. The local people, labors who are good for various forestry works, mostly reside in forest villages. They possess a good knowledge of the forest growth and indigenous medicinal plants.

Methodology

1. Initially important localities and diversity rich areas of wild medicinal plants were identified and demarcated, with the help of field survey. For studies the participation and involvement of tribal and local inhabitants were given prime importance. Potential habitats of important medicinal plants were identified.

Moreover potential threats to important habitats having high diversity of medicinal plants were listed and. Various collection and marketing methods of minor forest produce (MFPs) were observed in this area these were helpful in synthesizing information about current harvesting practices of medicinal plants both in the form of data and photographic record. During seasonal sample collection, enhno-botanical information was gathered form knowledge bearing persons of medicinal plants. Which includes some tribal and local people? Thereafter field notes were entered in the field diary and each specimen was given a specific collection number.

2. Herbarium of collected plant specimen was prepared following simultaneously the identification of plant specimens were carried out.
3. A list of all species found in the area was prepared. According to particular habit. A list of all species found in the areas was prepared keeping in view the IUCN list.

Table 9.1: List of Medicinal plants

Sl.No.	*Botanical Name*	*Comman Name*
1.	*Cordia macleodii* (Griff) Hook.	Silvat
2.	*Coccinia cordifolia* (L.)	Jangli-Kundru
3.	*Phyllanthus niruri* L.	Bhui-Aonla
4.	*Tephrosia purpuria* (L.) Pers.	Van-Kulthi H
5.	*Ophioglosum* sp.	Booti
6.	*Crotalaria ramosissima* Roxb.	Van San
7.	*Abrus precatorious* L.	Ratti
8.	*Cassia occidentalis* L.	Kasondhi
9.	*Richinus communies* L.	Arandi
10.	*Cassia tora* L.	Chakoda
11.	*Atylosia crassa* Dalz.	Vansemi
12.	*Celasturs paniculatus* Willd.	Orangul
13.	*Pavetta indica* L.	Narisa
14.	*Boswellia serrata* Roxb.	Salai
15.	*Smilax macrophylla* Roxb.	Ram-datun
16.	*Semecarpus anacardium* L.f.	Bhilwa
17.	*Pergularia daemia* (Forsk.) Chiov.	Dudhibel
18.	*Datura metal* L.	Dhatura kala
19.	*Crotalara sericea* Retz.	Van San
20.	*Jatropha curcus* L.	Ratanjot
21.	*Mitragyna parvifolia* (Roxb.) Korth.	Mundi
22.	*Madhuka latifolia* Roxb.	Mahua
23.	*Agave sislana* L.	Kataki

Contd...

Table 9.1–*Contd...*

Sl.No.	*Botanical Name*	*Comman Name*
24.	*Vallaris heyneri* Spreng.	DuddeBel
25.	*Buchanania lanzan* Spreng.	Char
26.	*Scheleichera oleosa* (Lour.) Oken	Kusum
27.	*Crotalaria verrucosa* L.	Hardul
28.	*Solanum melongena* L.	Banbhata
29.	*Colocasia indica* L.	Jangli-Arbi
30.	*Indigofera oblongifolia* Forak	–
31.	*Peristrophe bicalyculata* (Ritz.) Nccs	–
32.	*Antidesma diandrum* (Roxb.) Roth.	Khatta-amthi
33.	*Eulaliopsis binata* (Retz.) C.E. Hubb	Soom-Ghans
34.	*Pennisetum alopecureus* (Steud.)	Gangeua
35.	*Uraria picta* (Jacq) Desv. Ex. DC.	Patvan
36.	*Tectona grandis* L.F. Suppl.	Sagoun
37.	*Choloroxylon swietenia* DC.	Bhirra
38.	*Achyranthes aspera* L.	Chirchira
39.	*Curcuma caesia* Roxb.	Kalihaldi
40.	*Eranthemum purpurascens* Nees.	Ban-Tulsi
41.	*Centella asiatica* (L.) Urban.	Bramhi
42.	*Evolvulus alsinoides* L.	Shankh-puspi
43.	*Curcuma aromatica* L.	Van-Haldi
44.	*Globa bulbifera* Roxb.	Ganji
45.	*Zingiber cassumumar* Roxb.	Van-Adarak
46.	*Orozylum indicum* Vent.	San-padhar
47.	*Sonchus oleraceus* var. *asper*	–
48.	*Bambusa arumdinaceq* (Retz.) Willd	Katang-Bans
49.	*Lygodium* sp.	Hathajodi
50.	*Tephrosia villosa* (L.) Pers.	Van-Kulthi C
51.	*Desmodium heterocarpum* (L.) DC.	Charpatti
52.	*Pterocarpus marsupium* Schreb.	Bija
53.	*Litsea sebifera* Pers	Maida
54.	*Woodfordia fruticosa* (L.) Kurz.	Surtili
55.	*Soymaida febrifuga* (Roxb.) A.Juss	Rohan
56.	*Holarrhena antidysenterica* Wall	Kudo
57.	*Randia uliginosa* DC.	Safed-katul
58.	*Euphorbia hirta* L.	Doodhi
59.	*Diospyros melanoxyoln* Roxb.	Tendu

Results and Discussion

The result of intensive systematic medicinal and ethno medicinal observation was coveted throughout the district. Globally, there is an inversely interest of herbal drugs in the lively hood setups. Plants based drugs are safe and has no side affects. Due to this the demand of medicinal and aromatic plants in increasing day by day. This leads to unsustainable exploitation of natural resource by the growing population. There for it is essential to conserve these species before extinctions by way of their cultivation and stopping wide extraction form the forest area. Public awareness program should be intensified. Botanists should discharge their duty in building up the public opinion in favors of conservation of the biodiversity of medicinal plant of the district.

Acknowledgment

I am thankful to the forest department of Chhindwara (M.P.) district and also thankful to the principal of the institution and local tribal people of district.

References

Anand, Prakash and singh K.K. (2000). Observation on some high valued ethnomedicinal plants among the tribals of Uttar Pradesh, *J. Med. Arom.Pl.,*: 519-522.

Jain, S.K. (1991). The Dictionary of Indian Folk Medicines and Ethnobotany. Deep. Pub. New Delhi.

Kirtikar, K.R. and B.D. Basu (1918). Indian Medicinal Plants, 4 Vols. (2nd ed. Revised by Blatter, E. *et al.*, 1935, Allahabad.

Maheswari, J.K., Singh and S. Saha (1986). Ethonobotany of tribals of Mirazpur district, U.P., NBRI, Lucknow (UP).

Saini, D.C., S.K. Singh and S. Singh (1989). Hitherto unrecorded plants form upper Gangetic plain with its Ethnobotanical uses, *J. Bombay Nat. Hist. Soc.* 86(1): 118-119.

Satyawati, G.V., Gupta, A.K. and Tandon, n. (1987). Medicinal Plants of India Vol.11 312 pp. Indian Council of Medical Research, new Delhi, India.

Singh, H. (1988). Ethnobotaical treatment of piles by Bhoxas of Uttar Pradesh. *Ans. Sci. Life*. 8: 167-170.

Singh, K.K and Anand, Prakash (1996). Traditional medicinal plant therapy used for skincare by the tribals of Uttar Pradesh, *J.non-timber forest*, 3 (1/2): 51-55.

Singh, N.K. and D.P. Singh (2001). Ethnobotanical survey of District Balrampur, Flora and Fauna 7(2): 59-66.

Singh, V.P., M.K. Singh and R.C. Srivastava (2003). Medicinal flora of Jaunpur district (U.P.): *J. Econ. Tax. Bot.*

Srivastava C., H.N. Singh and R.C. Srivastava (2002). Medicinal plants of Gorakhpur district, *Proc. Nat. Symp. Env. Care and Sustain* 2: 211-250.

Srivastava, A.K. (2004). Some Threatened Medicinal Plants of Tarai Forest of Gorakpur of Uttar Pradesh. Role of Biotechnology in Medicinal and Arometc Plants vol. Xi, Ukaaz Publication. Hyderabad. pp 289-295.

Conservation of Biodiversity

Kritika J. Namdeo[1] and Swati Sahu[2]

[1]Department of Botany,
[2]Department of Zoology,
Govt. K. H. College, Abhanpur, C.G.

ABSTRACT

Biodiversity conservation is about saving life on earth in all its forms and keeping natural ecosystems functioning and healthy. It is estimated that the current species extinction rate is between 1,000 and 10,000 times higher than it would naturally be. The object of Biodiversity conservation is not to preserve forests because they are beautiful or wild or the habitat of wild animals; it is to ensure a steady ecosystem for nature's prosperity. The Convention on Biological Diversity (CBD) along with others advocates conservation of genes, species and ecosytems, but it is rarely, if ever, a practical possibility to operate at all three levels. Since it is hard to conserve all species, decisions have to be made about what species to prioritize. There are two different approaches: one that concentrates on identifying individual species of importance, i.e, SPECIES-BASED CONSERVATION and one that identifies important areas where it is hoped that actions will benefit a significant number of species, i.e., AREA-BASED CONSERVATION.

Conservation biology is faced with several controversial issues such as the dichotomy between the preservation of individual species versus a broader focus on the environment, the relative importance to give to endangered species. The conservation of whole communities emerges as the paramount strategy for maintaining the evolutionary potential of life on earth.

Keywords: *Biodiversity, Conservation, Species conservation, Forest conservation.*

Introduction

Biodiversity is the diversity of life on Earth. Biological diversity encompasses microorganism, plants, animals and ecosystems such as coral reefs, forests, rainforests, deserts etc. In the present era, human beings are the most dangerous cause of destruction of the earth's biodiversity. In 2006, the terms threatened, endangered or rare were used to describe the status of many species. According to the International Union for Conservation of Nature (IUCN), globally about one third of all known species are threatened with extinction. Even it is estimated that 25 per cent of all mammals will be extinct within 20 years and around 30 per cent of all species on earth will be extinct by 2050.

The object of Biodiversity conservation is not to preserve forests because they are beautiful or wild or the habitat of wild animals; it is to ensure a steady ecosystem for nature's prosperity. The Convention on Biological Diversity (CBD) along with others advocates conservation of genes, species and ecosytems, but it is rarely, if ever, a practical possibility to operate at all three levels. Since it is hard to conserve all species, decisions have to be made about what species to prioritize. There are two different approaches: one that concentrates on identifying individual species of importance, *i.e.*, SPECIES-BASED CONSERVATION and one that identifies important areas where it is hoped that actions will benefit a significant number of species, *i.e.*, AREA-BASED CONSERVATION.

Causes of Biodiversity Declines

The causes of biodiversity declines have been extensively explored in ecology. Some of the causes are as follows:

Influence of Natural Processes

Theoretical and empirical studies have identified a vast number of natural processes that can potentially disturb biodiversity. Such as overgrowth of a particular species in a community may disturb the ecological balance, natural calamities can extinct a species, climate change, overgrazing, intra and inter-specific competition and limited natural resources.

Influence of Human Activities

Human actions have resulted in multiple changes on a global scale that often drive contemporary biodiversity declines. In particular, land use changes, exotic species invasions, nutrient enrichment, and climate change are often considered some of the most ubiquitous and influential global ecosystem changes (Vitousek *et al.*, 1997, Chapin *et al.*, 2000, Benayas *et al.*, 2009, Butchart *et al.*, 2010). Unfortunately, the mechanisms by which global ecosystem changes influence biodiversity and ecosystem processes, and the combined effects of multiple changes, are often unclear. This greatly reduces the ability to predict future changes in biodiversity and ecosystem processes. Therefore, further investigation is needed to predict the consequences of global ecosystem changes.

Even if a small element of an ecosystem breaks down, the whole system's balance is threatened. Fresh water ecosystems are nowadays the most threatened ecosystems. **Invasive species** refer to those that would normally remain constrained from an ecosystem because of the presence of natural barriers. Since these barriers are no longer existing, invasive species invade the ecosystem, destroying native species. Human activities have been the major cause for encouraging invasive species.

Apparently stable areas of habitat may suffer from fragmentation, with significant impacts on their biodiversity. Fragmentation is caused by natural disturbance (such as fires or wind) or by land use change and habitat loss, such as the clearing of natural vegetation for agriculture or road construction, which divides previously continuous habitats. Larger remnants, and remnants that are close to other remnants, are less affected by fragmentation. Small fragments of habitat can only support small populations, which tend to be more vulnerable to extinction. Moreover, habitat along the edge of a fragment has a different climate and favors different species to the interior. Small fragments are therefore unfavorable for those species that require interior habitat, and they may lead to the extinction of those species. Species that are specialized to particular habitats and those whose dispersal abilities are weak suffer from fragmentation more than generalist species with good dispersal ability. Fragmentation affects all biomes, but especially forests and major freshwater systems.

Species can also be threatened by genetic pollution- uncontrolled hybridization and gene swamping. For instance, abundant species can interbreed with rare species thus causing swamping of the gene pool. Over exploitation is caused by activities such as over fishing, over hunting, excessive logging and illegal trade of wildlife. Over 25 per cent of global fisheries are being overfished at unsustainable levels.

Global warming is also becoming a major cause for loss of biodiversity. For example if the present rate of global warming continues, coral reefs which are biodiversity hotspots will disappear in 20-40 years. 10 per cent of all species might become extinct by 2015, if global warming continues.

Further, urbanization, industrialization, use of modern technology, deforestation, burning of forest to get land for multiple uses, habitat destruction, shifting cultivation, use of various chemicals and various types of pollution are also some of the causes of decline of biodiversity.

Consequences of Biodiversity Declines

Present-day threats are often multiple and of greater intensity than historical threats. The susceptibility of an ecological community to a given threat will depend on the events of the past that have shaped the current biota. If the current threats are novel, they will have dramatic effects on populations, since species will lack adaptations. Even if factors are similar to past factors (climate, for example, has always been variable to some degree), the intensity of some current-day factors is unprecedented (such as the rates and extent of habitat change). Furthermore, today's factors of extinction are often multiple—land use change, emerging disease, and invasive species are all occurring together, for instance. Because exposure to one threat type often makes a species more susceptible to a second, exposure to a second makes a species more susceptible to a third, and so on, consecutive, multiple threats to species may have unexpectedly dramatic impacts on biodiversity.

Each factor has a characteristic spatial and temporal scale at which it affects ecosystem services and human well-being. Climate change may operate on a spatial scale of a large region; political change may operate at the scale of a nation or a municipal district. Sociocultural change typically occurs slowly, on a time scale of decades, while economic forces tend to occur more rapidly. Because of the variability in ecosystems, their services, and human well-being in space and time, there may be mismatches or lags between the scale of the driver and the scale of its effects on ecosystem services.

The fate of declining species and habitats will depend on sources of inertia and the speed of their response to management interventions. Natural sources of inertia correspond to the time scales inherent to natural systems; for example, recovery of a population cannot proceed more quickly than the average turnover or generation time, and established recovery will often take several generations. On top of this is anthropogenic inertia resulting from the time scales inherent in human institutions for decision-making and implementation. For most systems, these two sources of inertia will lead to delays of years, and more often decades, in slowing and reversing a declining biodiversity trend. This analysis assumes that the factors of change could indeed be halted or reversed in the near term. Yet currently there is little evidence that any of the direct or indirect drivers are slowing or that any are well controlled at the large to global scale. More significantly, we have net yet seen all of the consequences of changes that occurred in the past.

The delay between a factor affecting a system and its consequences for biodiversity change can be highly variable. In the relatively well studied case of species extinctions, habitat loss is known to be a driver with particularly long lag times. In studies of tropical forest bird species the time from habitat fragmentation to species extinction has been estimated to have a half-life of decades to hundreds of years. Overall, these results suggest that about half of the species losses may occur over a period of 100 to 1,000 years. Therefore, humans have the opportunity to deploy active habitat restoration practices that may rescue some of the species that otherwise would have been in a trajectory toward extinction. Notwithstanding this, habitat restoration measures will not be likely to save the most sensitive species, which will become extinct soon after habitat loss.

Steps for Biodiversity Conservation

Theoretical and empirical studies have identified a vast number of natural processes that can potentially maintain biodiversity. Biodiversity can be maintained by moderately intense disturbances that reduce dominance by species that would otherwise competitively exclude subordinate species. For example, selective grazing by bison can promote plant diversity in grasslands (Collins *et al.*, 1998). Additionally, biodiversity can be maintained by resource partitioning, when species use different resources, or spatiotemporal partitioning, when species use the same resources at different times and places. For instance, plant species in the tundra can coexist by using different sources of nitrogen or use the same sources of nitrogen at different times of the growing season or at different soil depths (McKane *et al.*, 2002). Furthermore, biodiversity can be maintained by interspecific facilitation, which occurs when species positively influence one another by increasing the availability of limiting resources, or by decreasing the limiting effects of natural enemies or physical stresses. Although previous theoretical and empirical studies have identified numerous processes that can maintain biodiversity, ecologists and conservationists rarely know which of these mechanisms actually maintains biodiversity at any particular time and place. Thus, further investigation is needed to identify the natural processes that actually maintain biodiversity in intact ecosystems.

In some cases, human actions have promoted biodiversity. Conservation strategies, such as creating parks to protect biodiversity hotspots, have been effective but insufficient (Bruner *et al.*, 2001). For example, although biodiversity is often greater inside than outside parks, species extinctions continue. Similarly, restoration strategies, such as reinstating fire as a natural disturbance, have been effective but insufficient. Specifically, biodiversity and ecosystem services are greater in restored than in degraded ecosystems but lower in restored than in intact remnant ecosystems (Benayas *et al.*, 2009). Despite the positive effects of conservation and restoration efforts, biodiversity declines have not slowed (Butchart *et al.*, 2010). Thus, further investigation is needed to determine new conservation and restoration strategies.

Thus we can see that biodiversity which is crucial for the well being of life on earth, is coming under the threat of many factors related to human activities. There is an urgent need to take action to protect the magnificent biodiversity of our planet. We must create economic policies in order to maintain the Earth's biodiversity and take appropriate measures to protect habitats and species.

A synthesis across four ecological fields may increase our ability to understand, conserve, and restore ecosystems by providing a framework for considering the causes and consequences of biodiversity declines. **First**, maintenance of biodiversity research should be focused on the effects of natural processes on biodiversity. **Second**, biodiversity-stability research should be focused on the effects of biodiversity on various measures of stability. **Third**, biodiversity-ecosystem functioning research should be focused on the effects of biodiversity on ecosystem functioning and how this relationship mediates the effects of global ecosystem changes on human wellbeing. **Fourth**, global change ecology should be focused on the effects of global ecosystem changes on biodiversity, ecosystem functioning, and stability. Combining the relationships explored in each of these four fields will produces an inclusive framework and will help to understand the natural processes that promotes biodiversity, ecosystem functioning, and stability, further it will also help to understand the global ecosystem changes influencing ecosystems by altering the natural processes.

References

Benayas, J. M. R. *et al.*, Enhancement of biodiversity and ecosystem services by ecological restoration: A meta-analysis. *Science* **325**, 1121–1124 (2009).

Butchart, S. H. M. *et al.*, Global biodiversity: Indicators of recent declines. *Science* **328**, 1164–1168 (2010).

Chapin, F. S. *et al.*, Consequences of changing biodiversity. *Nature* **405**, 234–242 (2000).

Collins, S. L. *et al.*, Modulation of diversity by grazing and mowing in native tallgrass prairie. *Science* **280**, 745–747 (1998). 65.

Conner Bailey and Mike Skladany, "Aquacultural Development in Tropical Asia: A Re-evaluation," *Natural Resources Forum,* Vol. 15, No. 1 (1991), pp. 66-73.

Intergovernmental Panel on Climate Change (IPCC). *Climate Change 2007: Synthesis Report*. Geneva, 2007.

McKane, R. B. *et al.,* Resource-based niches provide a basis for plant species diversity and dominance in arctic tundra. *Nature* **415**, 68–71 (2002).

Vitousek, P.M., Mooney, H. A., and Lubchenco, J. *et al.,* 1997. Human domination of Earth's ecosystems. *Science* 277: 494-499.

11

Microalgae are Future Resource of Biofuel: A Review

Indu Soni[1], Preeti Tiwri[2] and Madhu Yamini Soni[3]

[1]Department of Botany,
Govt. K.K.B. College, Jetha, Janjgir-Champa, C.G.
[2]Principal,
Govt. R.L. College, Rajim, Gariaband, C.G.
[3]Department of Botany,
Govt. Dr. I.S. College, Akaltara, Janjgir-Champa, C.G.

ABSTRACT

The majority of algae that are intentionally cultivated fall into the category of microalgae. A micro alga is an ideal organism for biofuels. It can be cultivated on non-arable land and it grows in ponds or photo-bioreactors. Microalgae are capable of producing large amounts of biomass and usable oil in either high rate algal ponds or photo-bioreactor. This oil can then be turned into biodiesel which could be sold for use in automobiles. The species of Chlorella and Spirulena is more important for biofuel. Chlorella is known as 'Space Alga' present highly rich protein it is used for medicine and food. A small pill made from it is used by astronauts in space-ship. The present study can give information's on bio-fuel which produce from microalgae. Advantages of microalgae from biodiesel is high productivity and contain high oil content compared with terrestrial crops, up to 77 per cent lipid content, negligible lignocellulosic biomass component, diesel more versatile and energy dense than bio-ethanol. Terrestrial oil crops are unable to meet in future bio-oil demand but by using of micro-algae it is possible.

Keywords: *Microalgae, Biofuel, Photo bioreactor, Space-alga, Bio-ethanol.*

Introduction

Algae are a diverse group of aquatic and marine organisms. Like plants, algae carry out photosynthesis, using sunlight to convert carbon dioxide, water and nutrients to oxygen and biomass (carbohydrates, oils and proteins). Micro algae can provide a variety of fuels for transport, heating or electricity generation, including biodiesel, aviation fuel and biogas. Algal bio-fuels are at an experimental stage and are more expensive than fossil fuels. Micro algae are cultivated in a variety of aqueous systems, from open air ponds to closed photo bioreactors with closely controlled environments. **(Graham, L. E.; *et al.*, 2008), Lehr, *et al.* (2009), Shen, Y.; *et al.*, 2009).** Algae are the most potentially significant sources of sustainable biofuels in the future of renewable energy. They play role in the development of the earth's biosphere was of unique importance.

Algae-based biofuel production has a number of potential advantages:

- ☆ Biofuels and by products can be synthesized from a large variety of algae.
- ☆ Algae have a rapid growth rate.
- ☆ Algae can be cultivated in brackish coastal water and seawater.
- ☆ Some land areas that are unsuitable for agricultural can be used to cultivate algae.
- ☆ Algae nutrient uptake uses high nitrogen, silicon, phosphate, and sulfate nutrients from human or animal waste.
- ☆ Algae can sequester carbon dioxide (CO_2) from industrial sources.

Algae's are define in two type's- first is microalgae and second is macro-algae.

Microalgae are single celled algae and it grown in open ponds or in enclosed systems known as photobioreactors. Many microalgae are produce biofuels. Larger species of algae, including seaweed, are known as macro-algae, and can be grown at sea or in tanks.

Microalgae

It is widely reported that microalgae have advantages over first-generation biofuel crops, including higher yields, faster growth and lower requirements for land. Significant investment has been made in the development of algal biofuels, particularly within the energy and aviation sectors. Algal biofuels could become more competitive as fossil fuel prices increase, although this would also increase the cost of algal fuel production through increases in the cost of building materials, energy or nutrients. Microalgae hold great potential as a renewable source of oil for biofuel production.

Macroalgae

Macroalgae are currently used mainly for food. The cost of production is high-recent estimates suggest that biogas from seaweed could be 7-15 times more expensive than natural gas.

Biofuel Production

The potential for microalgae to generate biofuel is augmented by their fast growth, reaching maturity in as little as three days while producing in excess of half their weight in oil (Demirbas, M.F. 2011). From 1978 to 1996 the U.S. Department of Energy funded a program to develop renewable transportation fuels from algae. Microalgae produce much higher yields of fuel-producing biomass than other traditional fuel feed stocks and it doesn't compete with food crops. High productivity and high lipid content make algae a potentially important future source of biofuel (Bajpai, D., Tyagi, V.K. 2006). Algae stored energy in the form of carbohydrate and oil which joined with their high productivity. Micro algae can be produced 2,000 to 5,000 Galan/hectare/year of biofuel.

Oil Content from Few Microalgae Species

Microalgal Species	*Oil Content (Dry weight)*
Ankistrodesmus TR-87	28-40
Botryococcus braunii	29-75
Chlorella sp.	29
Dunaliella tertiolecta	36-42
Nannochloris	31(6-63)
Nannochloropsis	46(31-68)
Phaeodactylum tricornutum	31
Scenedesmus TR-84	45
Stichococcus	33(9-59)
Tetraselmis suecica	15-32
Thalassiosira pseudonana	(21-31)
Crypthecodinium cohnii	20
Neochloris oleoabundans	35-54
Schiochytrium	50-77

Source: This has been derived from the table of 'Marc Y. Menetrez' (2012) paper.

Fuels obtained from Algae

The types of fuel that can be derive from algal species.

Biodiesel is discovered from 1978 to 1996 the U.S. Department of Energy funded a program to develop renewable transportation fuels from algae. It is produced from the oil (lipid) content of microalgae by chemical processing to give biodiesel (fatty acid methyl ester; FAME) and glycerol, used in chemical manufacturing blends with up to 5 per cent FAME are common. It is an alternative fuel that can be produced from a variety of renewable sources. These sources include canola oil, soybean oil, sunflower oil, cottonseed oil, animal fats, and lipids produced by algae. Algal biodiesel is a good replacement for standard crop biodiesels.

Methods

The following common method are used for producing of microalgae:

- ☆ **Biodiesel harvesting-:** Algae can be harvested using micro-screens, by centrifugation, by flocculation and by froth flotation. Algal-oil processes into biodiesel as easily as oil derived from land-based crops. Algae can be harvested in open-ponds and photo bioreactors.

Open-pond System

- ☆ **Open pond**-: Ponds and lakes are open to the elements. Open ponds are highly vulnerable to contamination by other microorganisms, such as other algal species or bacteria. Open systems also do not offer control over temperature and lighting. The growing season is largely dependent on location and, aside from tropical areas, is limited to the warmer months. Open pond systems are used for the majority of algae cultivation, especially those strains with high oil content (Briggs, *et al.* (2004; Gonzalez-Fernandez, *et al.*, 2011; Venkata Mohan, *et al.*, 2007).

Algal Culture by Open Pond.

Photo Bioreactors

Algae can also be grown in a photo bioreactor (PBR). A PBR is a bioreactor which incorporates a light source. Virtually any translucent container could be called a PBR; however the term is more commonly used to define a closed system, as opposed to an open tank or pond because PBR systems are closed, the cultivator must provide all nutrients, including CO_2.

- ☆ **Bioethenol** is used as a substitute for petrol, and is produced by fermentation of the carbohydrade content of algae by microbes or yeast. Ethanol production from algae has very interesting prospects in future.
- ☆ **Hydrocarban** can be produced by treating unprocessed algae with high pressures and temperatures or by chemical conversion of micro-algal oil. Hydrocarbons are used as aviation fuel.

Photo-bioreactors.

- ☆ **Hydrogen** discovered first in 1939 by Hans Gaffrom. Hydrogen is produced from water by some species of algae and bacteria in the absence of oxygen. Hydrogen can be used to produce heat, electricity,or power (via fuel cells) for transportation. The generation of hydrogen was shown to be dependent on the growth media composition, culture conditions, and PBR design. An additional biofuel, for example, comes from the microalga *Chlamydomonas reinhardtii*, which has been demonstrated to grow in the laboratory to produce hydrogen (Beer, *et al.*, 2009; Mullner, (2007).
- ☆ **Biomass**–Micro-algae can be grown to produce biomass. It burned to produce heat and electricity. It can still produce greenhouse gases. Production of microalgal biomass for making biodiesel has been extensively evaluated in raceway ponds in studies sponsored by the United States Department of Energy (Sheehan *et al.*, 1998).

Economic Importance of Micro Algae

Microalgae lipid production has the greatest potential for the production of renewable fuels. Although many processes (with theoretical costs) use various forms of microalgae, information on algal bio-fuel production costs is limited to biodiesel. Microalgae can produce oil and protein in higher mass and that is used to make biofuel or feedstock. Micro-algal oil is converted in to fuel and residue algal biomass is dried pelletized and then it is used in industrial boiler and other source of electricity production in form of fuel. The fluctuating price of petroleum continues to set the economic standard that all bio-fuels must meet to be competitive. Biodiesel production can be achieved by both heterotrophic and autotrophic process. The cost of feedstock or carbon source (carbohydrate) in the heterotrophic process account for 60-75 per cent of the total cost biodiesel (Brennan, L. and Owende, P. 2010; Borowitzka, M.A. 1997).

Conclusion

Algae-derived biofuel will directly impact the generation of transportation fuels (biodiesel, ethanol, and petroleum), and as part of the future of renewable fuel it will also impact many environmental and economic resources. The rapid commercial expansion of the algae biofuels industry is an excellent example of sustainable product development with dramatic future potential for contributions to fuel supplies, yet many questions regarding algae production remain unanswered. It has a useful and safe purpose for the economic feasibility and environmental sustainability of the process. Photo-bioreactors provide a controlled environment that can be tailored to the specific demands of highly productive microalgae to attain a consistently good annual yield of oil. Today petrol and diesel are very costly and in this situation we have to find out alternative fuel based on petroleum. Terrestrial oil crops are unable to meet in future bio-oil demand but by using of micro-algae it is possible. So we can called micro algae is a future green gold for biofuel.

References

Borowitzka, M.A. (1997). Microalgae for aquaculture: opportunities and constraints. *J. Appl. Phycl.* **9:** 393-401.

Brennan, L. and Owende, P. (2010). Biofules from microalgae- A review of technologies for production, processing and extraction of biofules and co-products. *Renewable sustainable energy* **14:** 557-577.

Bajpai, D.; Tyagi, V. K. (2006). Biodiesel: Source, production, composition, properties and its benefits. *J. Oleo Sci.* **55:** 487– 502.

Beer, L. L.; Boyd, E. S.; Peters, J. W.; Posewitz, M. C. (2009). Engineering algae for biohydrogen and biofuel production. *Curr. Opin. Biotechnol.* **20:** 264–271.

Demirbas, M. F. (2001). Biofuels from algae for sustainable development. *Appl. Energy,* **88:** 3473–3480.

Graham, L. E.; Graham, J. E.; Wilcox, L. W. (2008). Algae, 2nd ed.; Benjamin-Cummings Publishing: *Menlo Park, CA,* **ISBN:** 0321559657.

Gonzaìlez-Fernaìndez, C.; Molinuevo-Salces, B.; García- Gonzaìlez, M.C. (2011). Nitrogen transformations under different conditions in open ponds by means of microalgae–bacteria consortium treating pig slurry. *Bioresour. Technol.* **102:** 960–966.

Lehr, F. Posten, (2009). Closed photo-bioreactors as tools for biofuel production. *Curr. Opin. Biotechnol.* **20:** 280–285.

Mullner, K.; Happe, T. (2007). Biofuel from algae-photobiological hydrogen production and CO2-fixation. *Int. J. Energy Technol. Policy.* **5:** 290–295.

Marc, Y. Menetrez (2012). An overview of algae bioful production and potential environmental impact. *Env. Sci. Tec.* **46:** 7073-7085.

Shen, Y.; Pei, Z.; Yuan, W.; Mao, E (2009). Effect of nitrogen and extraction method on algae lipid yield. *Int. J. Agric. Biol. Eng.* **2 (1):** 51–57.

Sheehan J, Dunahay T, Benemann J, Roessler P. (1998). A look back at the U.S. Department of Energy's Aquatic Species Program-biodiesel from algae. *National Renewable Energy Laboratory, Golden, CO;. Report NREL/TP*-580–24190.

Venkata Mohan, S.; Bhaskar, Y. B.; Krishna, T. M.; Chandrasekhara Rao, N.; Lalit Babu, V.; Sarma, P. N. (2007). Biohydrogen production from chemical wastewater as substrate by selectively enriched anaerobic mixed consortia: Influence of fermentation pH and substrate composition. *Int. J. Hydrogen Energy.* **32**: 2286–2295.

12

Study on Antimicrobial Activity of different Solvent Extracts of *Ocimum sanctum* Linn. against some Bacterial Species

Prerna Soni[1] and Shobha Gawri[2]

[1]Department of Biotechnology,
[2]Principal,
Seth Phoolchand Agrawal Smrity Mahavidyalaya Nawapara, Raipur

ABSTRACT

Antimicrobial activity of medicinal plants has formed the basis of many applications. Medicinal plants have attracted a great deal of scientific interest due to their potential as a source of natural biologically active compounds. The present study was to evaluate the qualitative estimation of phytochemicals and antimicrobial activity of aqueous and solvent extracts of leaves, stem and roots of Ocimum sanctum against pathogenic bacteria i.e. Staphyllococcus aureus, Escherichia coli, and Bacillus subtilis. Significant antimicrobial activity of the plant extracts was quantitatively assessed by the presence of zone diameter of inhibition at three different concentrations of 50µg/ml, 100µg/ml and 200µg/ml. Largest zone of inhibition were observed at 200µg/ml concentration of all extracts. The qualitative phytochemical study has been shown the presence of steroids, flavonoids, alkaloids, phenols and reducing sugars suggested antimicrobial properties that can be used in new drugs for the therapy of infectious diseases.

***Keywords**: Ocimum sanctum, S. aureus, E. coli, B. Subtilis, Antimicrobial.*

Introduction

Herbal medicine is one of the most remarkable uses of plant based biodiversity. Medicinal plant plays a key role in world healthcare system. Ethno botanical information from India estimates that more than 6000 higher plant species forming about 40 per cent of the higher plant diversity are used in its codified and folk healthcare traditions [23]. This situation has interest for searching new antimicrobials from natural sophisticated traditional medicine system that have given rise to some important drugs still in use today [4 and 6]. Plant metabolites and plant based drugs appear to be one of the better alternatives as they are known to have minimal toxicity and cost effective in contrast to synthetic agents.

Ocimum species often referred to as the "king of the herb. *Ocimum* is a genus of about 35 species of aromatic annual and perennial herbs and shrubs in the family Lamiaceae, mostly native to the tropical and warm temperate regions of the Old World. According to World Health Organization [19], medicinal plants would be the best source to obtain a variety of drugs. Therefore, such plants should be investigated to better understand their properties, safety and efficiency [7].

Tulsi has also been recognized by the rishis for thousands of years as a prime herb in Ayurvedic treatment. Tulsi is abundant in essential oils and antioxidants, which are tremendously effective in reducing various types of diseases. Therefore the aim of present study was to evaluate the antimicrobial activity of extracts from the different parts *Ocimum sanctum* plant against three pathogenic bacteria *Streptococcus aureus, Escherichia coli* and *Bacillus subtilis*.

Plant Description

The studied plant *Ocimum sanctum* L. known as Krishna tulsi in Hindi and holy basil in English, is an annual shrub that has a height of 2 to 4 feet found throughout India. The whole plant are of purple colour. Leaves are 1 to 2 inch long. It has a bud at the apex of the petiole. Flowers are small and of purple colour. Seeds are oval and

flattened having brown colour and bears black spots. The main chemical constituents are Oleanolic acid, Ursolic acid, Rosmarinic acid, Eugenol, Carvacrol, Linalool, and β-caryophyllene has been used extensively for many years as therapeutic agent. It shows a number of medicinal activities and used in anticancer, antifertility, antidiabetic and various other disease [17].

Materials and Methods

Collection of Plant Material

The different plant parts of *Ocimum sanctum* L. were collected from, Kabir Aashram, Nawapara, district, Raipur during December 2015. It was authenticated by Dr. U. Tewari, Department of Botany, Govt. Science College, Bilaspur,Chhattisgarh. The leaves, root and stem the plant were washed thoroughly 2-3 times with running water and one with sterile distilled water. It was then air dried on sterile blotter and shade.

Preparation of Extract

The plant leaves were washed with deionised water and disinfected with 0.1 per cent HgCl solution for 5min and dried in shade for 15 days. The dried materials were ground to fine powder with the help of electrical grinder [5,10].

Aqueous Extraction

20 gm. of air-dried powder of leaves and root was added to distilled water and boiled on slow heat for 24 h. It was then filtered through filter paper. The supernatant was collected. This procedure was repeated twice. The supernatant concentrated to make the final volume one-fourth of the original volume with the help of water bath [15].

Methanol Extraction

20 gm., of air-dried powder was taken in 100 ml of methanol in a conical flask, plugged with cotton wool and then kept on a rotary shaker at 150 rpm for 24 h. After 24 hours the supernatant was collected and the solvent was evaporated to make the final volume one fourth of the original volume with the help of water bath and stored at 4°C in airtight bottles [24].

Ethanol Extraction

20 gm., of air-dried powder was taken in 100 ml of methanol in a conical flask, plugged with cotton wool and then kept on a rotary shaker at 150 rpm for 24 h. After 24 hours the supernatant was collected and the solvent was evaporated to make the final volume one fourth of the original volume with the help of water bath and stored at 4°C in airtight bottles [24].

Test Microorganisms

The bacterial strains studied are *Staphylococcus aureus, Escherichia coli* and *Bacillus subtilis.* Microorganisms were maintained at 4 C on nutrient agar slants. Each of the microorganisms was freshly cultured prior to susceptibility testing by transferring them into a separate sterile test tube containing nutrient broth and incubated overnight at 37 C. A microbial loop was used to remove a colony of each bacterium from pure culture and transfer it into nutrient broth.

Preparation of Inoculum

Each organism was recovered for testing by sub culturing on fresh media. A loop ful inoculum of each bacterium was suspended in 5ml of nutrient broth and incubated overnight at 37 C. These overnight cultures were used for the study.

Preparation of Media

The growth media employed in the present study included nutrient agar media: Peptone-5.0g; Beef extract- 3.0g; Sodium chloride-5.0g:Agar-15.0g;Distilled water-1000ml. Nutrient broth is composed of without agar. The medium was adjusted to pH 7.4 and sterilized by autoclaving at 15 lbs pressure and 121 C temperature for 15 min.

Disc Diffusion Method

Antimicrobial activity of the leaf, stem and root extracts of *Ocimum sanctum* was tested by using the disc diffusion method [12]. Sterile nutrient agar plates were prepared and bacterial strains inoculated by spread plate methods in aseptic conditions. The sterile filter paper disc of 5mm diameter (Whatman no. 1 filter paper) was prepared. Also methanolic, ethanolic and aqueous extracts were prepared in various concentrations of 50 µg/ml and 100µg/ml and 200 µg/ml experiment disc were dipped for before putting on medium. The sterile impregnated disc with plant extracts were placed on the agar surface with framed forceps and gently pressed. Filter paper discs soaked in solvent were used for negative controls. All the plates were incubated at 37 C for 24 hours. After incubation, the size (diameter) of the inhibition zones was measured by antibiotic zone scale meter.

Results and Discussion

When tested by the disc diffusion method the methanol leaf extract of *Ocimum sanctum* showed significant higher activity around maximum range between 6.5- 7.3 mm (Figure 12.1) comparatively with ethanol extracts which showed antimicrobial activity range between 5.7- 6.8mm with all tested bacterial culture (Figure 12.2). As the concentration of leaf, stem and root extracts of plant increased from 50< 100 < 200, the zone of inhibition with different microbial cultures were also increased and maximum zone of inhibition from 7.3mm were reported in 200 µg/ml concentration of root extract with methanol, whereas the least values were obtained with the 50µg/ml concentration of leaf extract that was between in ranges from 2.1mm.

Several reports have been shown that bioactive compounds isolated from plant extracts have inhibitory effect on pathogens strains. [14, 13, 1, 20]. Medicinal plant extracts of *Withania somanifera* with ethanol and methanol posses potential antimicrobial activity against *Staphylococcus aureus, Bacillus subtilis, Proteus vulgaris* and *Pseudomonas aeuroginosa* [21].

S.No.	Used Plant Parts	Concentration (in µg/ml)		
		50	100	200
		Zone of inhibition (in mm)		
1.	Root	3.1	5.6	7.3
2.	Stem	2.2	4.3	6.2
3.	Leaf	2.1	3.8	6.5

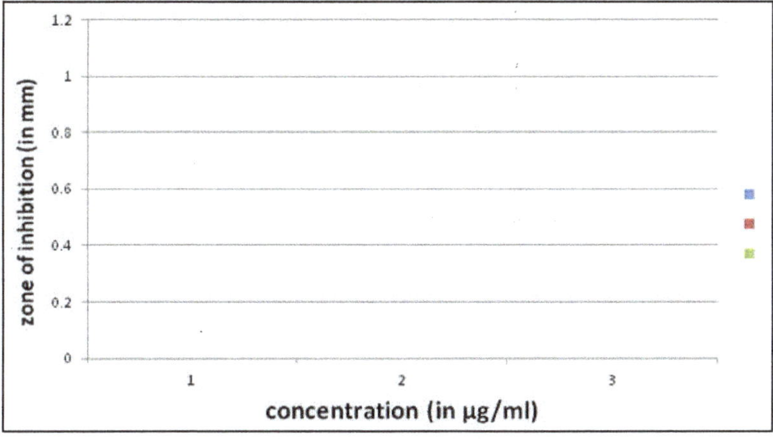

Figure 12.1: Inhibitory Zone of Methanolic Extracted Plant Parts against different Bacterial Species.

S.No.	Used Plant Parts	Concentration (in µg/ml)		
		50	100	200
		Zone of inhibition (in mm)		
1.	Root	3.4	5.7	6.8
2.	Stem	3.0	4.3	5.7
3.	Leaf	2.6	4.8	6.3

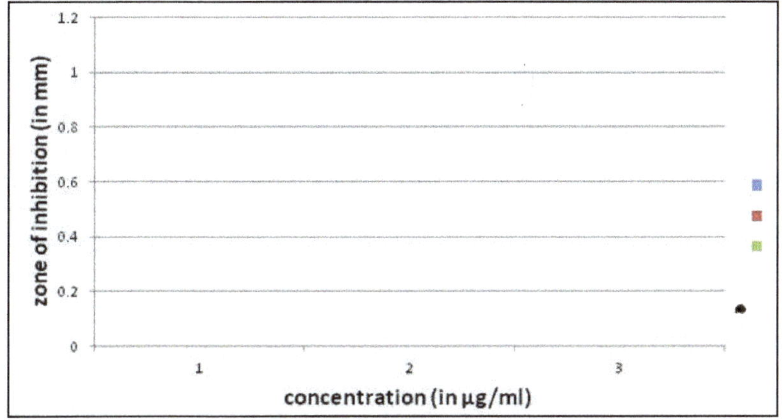

Figure 12.2: Inhibitory Zone of Ethanolic Extracted Plant Parts against different Bacterial Species.

S.No.	Used Plant Parts	Concentration (in µg/ml)		
		50	100	200
		Zone of inhibition (in mm)		
1.	Root	2.5	4.9	6.7
2.	Stem	2.4	4.8	5.8
3.	Leaf	2.6	4.9	6.3

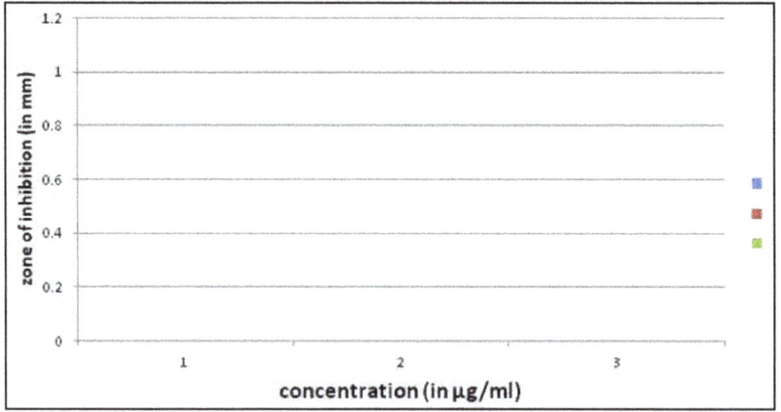

Figure 12.3: Inhibitory Zone of Aqueous Extracted Plant Parts against different Bacterial Species.

Many reports are available on the antiviral, antibacterial, antifungal, anthelmintic and anti-inflammatory properties of plants (Samy and Ignacimuthu, 2000; Palombo *et al.*, 2001; Kumaraswamy *et al.*, 2002; Stepanovic *et al.*, 2003; Behra and Misra 2005; Bylka *et al.*, 2004; Govindarajan *et al.*, 2006; Kambizi and Afolayan 2008). The results of present study supports the traditional usage of plant and *Ocimum sanctum* plant extracts which posses compounds with antibacterial properties that can be used as antibacterial agents in new drugs for the therapy of infectious diseases caused by pathogens and further work may be carried out for pharmacological evaluation.

Conclusion

The present study is related to pharmacognostical and preliminary phytochemical screening of *Ocimum sanctum* leaves provided useful information about its correct identity and evaluation. Phytochemical study is also useful to isolate the pharmacologically active principles present in the drug. All the above findings support the traditional knowledge of local healer and it is a preliminary, scientific validation for the use of these plants for antibacterial activity to promote proper conservations and sustainable use of such plant resources.

Acknowledgement

The author wish to express their gratitude to the department of biotechnology, Seth Phool Chand Agrawal Smriti College, Nawapara dist.-Raipur for providing laboratory facility.

References

1. Balasaraswathi R., Sadasivam S., Ward M. and Walker J.M. (1998). An antiviral protein from *Bouganvillea spectabilis* roots, purification and characterization. *Phytochem.*47(8):1561- 1565.
2. Behera S.K. and M.K. Mishra, (2005). Indigenous phytotherapy for genitor-urinary diseases used by the Kandha tribe of Orissa, *India. J. ethnopharmacol.*, 102:319-325.
3. Bylka W., M. Szaufer, Hajdrych I. Matalawskan and O. Goslinka (2004). Antimicrobial activity of isocytisoside and extracts of *Aquilegia vulgaris* L. Lett. *Appl. Microbiol.*, 39:93-97.
4. Cowan M.M.(1999). Plant products as antimicrobial agents, *Clinical Microbiol. Rev.*, 12, 564-582.
5. Dandapat S., M. Kumar, A. Kumar and M.P. Sinha (2013b). Therapeutic efficacy and nutritional potentiality of Indian Bay leaf (*Cinamomum tamala* Buch.-Hem.), *International Journal of Pharmacy*, 3(4): 779-785.
6. De Smet(1997).P.A.G.M. The role of plant derived drugs and herbal medicines in healthcare. Drugs.54, 801-840.
7. Ellof J.N.(1998). *Journal of Ethnopharmacolgy*, 60, 1-6.
8. Govindarajan, R., M. Vijayakumar, M. Singh, C. H. V. Rao, A. Shirwaikar, A. K. S. Rawat and P. Pushpangadan (2006). Antiulcer and antimicrobial activity of *Anogeissus latifolia*. *J. Ethnopharmacol.*, 106:57-61.
9. Kambizi L. and A.J.Afolayam (2008). Extracts from *Aloe ferox* and *Withania sonmnifera* inhibit *Candida albicans* and *Neisseria gonorrhea*. *African J. Biotechnol.*, 7:12-15.
10. Kumar M., A. Kumar, S. Dandapat and M.P. Sinha (2013b). Growth inhibitory impact of *Adhatoda vasica* and *Vitex negundo* on some human pathogens, The Ecoscan, 4(special issue): 241-245.
11. Kumaraswamy Y., P.J. Cox, M. Jaspars, L. Nahar and S.D. Sarker (2002). Screening seeds of Scottish plants for antibacterial actrivity. *J. Ethnopharmacol.*, 83:73-77.
12. Mahmood K., Yaqoob U. and Bajwa R. (2008). Antibacterial activity of essential oil of *Ocimum sanctum* L., *Mycopath.* 6: 63-65.
13. Meng,J.C., Zhu, Q.X.and Tan, R.X. (2000). New antimicrobial and sesquiterpenes from *Soroseris hookeriana* subsp.erysimoides. *Planta Medica*,.66, 541-544.
14. Mimica-Dukic N., Kujundzic S., Sokovic M, and Couladis M (2003). Essential oil composition and antifungal activity of *Foeniculum vulgare* Mill obtained by different distillation conditions. *Phytother Res.*17 (4):368-71.
15. Owolade B.F.and YOK Osikanlu (1999). Journal of Sustainable Agriculture and Environment, 1[2], 198-202.

16. Palombo E.A. and S.J. Semple (2001). Antibacterial activity of traditional medicinal plants. *J. Ethnopharmacol.,* 77:151-157.

17. Prakash P. and Neelu gupta (2005). Therapeutic uses of *Ocimum sanctum* with a note on eugenol and its pharmacological uses. *Indian j physiol pharmacol,* 49(2):125-131.

18. Samy R.P. and S. Ignacimuthu (2000). Antibacterial activity of of some folklore medicinal plants used by tribals in Western Ghats in India. *J. Ethnopharmacol.,* 69:63-71.

19. Santos PVR, ACX Oliveira, TCB Tomassini (1995). *Rev. Farm. Bioquím.,* 31, 35-38.

20. Sharma A., Meena S., and Barman N.(2011). Efficacy of ethyle acetate and ether extract of *Terminalia chebula* Retz against some human pathogenic strains. *International Journal of Pharmatech Research.* 3(2):724-727.

21. Soni P., A.N. Bahadur, U.Tewari(2015). Study on antimicrobial activity of leaf extract of *Withania somnifera* L. Dunal against clinical pathogens. *Int. J. Adv. Res. Biol. Sci.* 2(11):193–196.

22. Stepanovic S., N. Antic, I. Dakic and M. Svabicvlahovic (2003). *In vitro* antimicrobial activity of propilis and antimicrobial drugs. *Microbiol. Res.,* 158:353-357.

23. Ved D.K.,and G.S. Goraya (2007). Demand and Supply of Medicinal Plants in India National Medicinal Plant Board, New Delhi and FRLHT, Bangalore, India.

24. W.H.O.(1983a). Protocol CG-04 preparation of alcoholic extract for bioassay and phytochemical studies [APJF/IP, 100 1A], Geneva.

13

Phytochemical Analysis of *Mimosa hamata* Linn.

***Bhawana Pandey*[1] *and Sudha Agrawal*[2]**

[1]*Department of Microbiology and Biotechnology,*
[2]*Department of Zoology,*
Bhilai Mahila Mahavidyalaya,
Hospital Sector, Bhilai, Distt. Durg, Chhattisgarh

ABSTRACT

Ethanolic extracts of Mimosa hamata leaves were screened for phytochemical constituents. Phytochemical analysis of the extract revealed that the antimicrobial activity of the plant materials is due to the presence of active constituents like alkaloids or tannins.

Mimosa hamata is used in disease related to blood and bile, bilious fever, piles, jaundice, leprosy, ulcer and smallpox. In the present study ethanolic extracts of Mimosa hamata leaves and roots sample were obtained using soxhlet apparatus. Phytochemical studies for the presence of revealed that tannin and proteins are present in both the samples.

Keywords: *Mimosa hamata, Antimicrobial activity, Phytochemical.*

Introduction

Kingdom: Angiospers

Order: Fabales

Family: Fabaceae

Genus: *Mimosa*

Species: *M. hamata*

Mimosa hamata Family Mimosae known as Gulabi Babool plant. It is a flowering shrub in pea family. The plant is distributed through out in India desert locality. A diffuse prickly under shrub, is about 45-90 cm in height. Leaves bipinnately compound, pinnate 2-4 delicately arranged with 10-20 pairs of leaflets, rachis clothed with ascending bristles. Flowers pink, in globose heads, penduncles prickly, usually in auxiliary pairs all along the branches. Fruits bristly pods, flat, straw colored consisting of 3-5 one seeded segments. The roots and leaves are commonly used for camrl and goat fodder and soil binder. It is also used as blood purifier, in snake bite, cough, weakness, leprosy and in headache. (Vaidyaratanm, 2001).The present study intends to study about the phyto constituents of the plant extracts of *Mimosa hamata* against pathogenic microbes.

Figure 13.1: *Mimosa hamata* Plant.

Figure 13.2: Powdered *Mimosa hamata* Leaves.

Many plants species used traditionally have potential antimicrobial and antiviral properties (Shelef *et al.*, 1983) and this has raised the optimistic thinking of scientists about the future of phyto-antimicrobial agents. (Das *et al.*, 1999). Mimosa plant has a history of use for the treatment of various ailments and the most commonly used plant part for this purpose is the root, but flowers bark and fruit can also be utilized.

Several research works have been carried out to study about the phytochemical components of *Mimosa hamata* (Ahmad *et al.*, 2001; Arthur, 1954) and also about the antimicrobial activity of the plant (Palacios *et al.*, 1991). The major chemical substances of interest in these surveys were the alkaloids and steroidal saponins, however also been reported (Lozoya and Lozaya, 1989). The methanolic extract of leaves of *Mimosa* showed the presence of bioactive components like terpenoids, flavonoids, glycosides, alkaloids, quinines, phenols, tannins, saponins and coumarin (Gandhiraja *et al.*, 2009). In Manipur, a state in India, it is reported that the consumption of the decoction of leaves boiled in water causes diuresis, and is used in urinary tract infection. This plant has hepatoprotective, hypolipidemic, antifertility, antihapatotoxic, anti convulsant, anti depressant and wound healing properties. The seeds of the plant was also said to have diuretic property (Krishnaraju *et al.*, 2006). Roots of mimosa contain tannin, ash, calcium oxalate crystals and alkaloid mimosine (Oudhia *et al.*, 2006).

This plant has a history of use for the treatment of various ailments and the most commonly used plant part for this purpose is the root, but flowers, bark and fruit can also be utilized. Several research works have been carried out to study about the Phytochemical components of *Mimosa pudica* (Ahmad and Beg, 2001; Arthur, 1954; Deininger, 1984) and also about the antimicrobial activity of the plant (Palacios *et al.*, 1991; Ojalaa *et al.*, 1999, Pandey *et al.*, 2015) The present study intends to study about the phytochemicals in plant extracts of *Mimosa hamata*.

Collection of Plant Materials

Fresh leaves and root of *Mimosa hamata* were collected from Hydrabad.

Sample Preparation

The sample leaf and root were washed with sterile water, shade dried, powdered and kept in an air tight container for further use.

About 20g of the powdered leaves were soaked in 100ml of methanol. It was left for 24 hours so that alkaloids, terpenoids and other constituents if present get dissolved. The methanolic extract was filtered using Whatmann 41 filter paper. It was again filtered through Sodium sulphate in order to remove the traces of moisture.

Plant Extraction Method

Extraction

20 gms of each sample were taken and extracted separately with 250 ml ethanol using soxhlet apparatus. The extract were collected and dried. The condensed extract was then dissolved in ethanol to the concentration of 100mg/ml. After that allow for 5 cycles and switch of the apparatus and then take the sample solution and extracted solution in a beaker and cover it with a paper and make holes on the paper for the evaporation of the solvent.Allow it for drying and then collect the residue from the beaker.

Phytochemical Screening (Dey and Raman, 1957)

Phytochemical screening of the plant extract was carried out as per the methods and tests given by Dey and Raman (1957) to decipher the presence or absence of various phytocompounds. The stock concentration of plant extract 10 mg/ml was used.

Test for Tannins

Preparation of 0.1 per cent Ferric Chloride

To 99.9 ml of distilled water 0.1ml of ferric chloride reagent was added.

Ferric Chloride Test

1 ml of the sample taken and a few drops of 0.1 per cent ferric chloride was added and observed for brownish green or blue, black colouration.

Test for Saponins

To 1 ml of extract 5 ml of distilled water was added and shaken vigorously. Observed for soaking appearance indicates the presence of saponins.

Test for Flavonoids

To 1 ml of extract 5 ml of dilute ammonia solution was added, followed by addition of concentrated sulphuric acid along the sides of the tube. Appearance of yellow colouration.

Test for Alkaloids

1 ml of sample was taken to that few drops of Dragandoff reagent was added and observed for orange red colour.

Test for Protein

1 ml of sample was taken to that few drops of Bradford reagent was added. The blue colour was observed.

Test for Steroids

1 ml of the filtrate was taken to that 10 per cent concentration H_2SO_4 was added and observed for green colour.

Test for Anthroquinones

1 ml of sample was taken to that aqueous ammonia (shaking) was added and observed for change in colour of aqueous layer (Pink, Red or Violet).

Results and Discussion

Phytochemical Analysis

The crude extract of both samples were studied and the result were tabulated (Table 13.1) Phytochemical, which process many Ecological and physiological roles as widely distributed as plant constituents. Phytochemical exhibit wide range of biological effects as constituents at their antioxidant properties. The phytochemical analysis of the crude extract indicated the presence of tannins, proteins and steroids.

These compounds are known to be biological active and therefore aid the antimicrobial activity. Tannins have been found to form irreversible complexes with highly rich protein resulting in the inhibition of cell protein synthesis.

Tannins are known to react with protein to provide difficult tanning effect which is important for the treatment of influenced or ulcerated tissues. Herbs that that have tannins have the main component astringen are used for treating intestinal disorder such as diarrhea and dysentery. The presence of tannin in *Mimosa hamata* is the traditional treatment for ailments.

Steroidal compounds present in *Mimosa hamata* extracts are important due to their relationship with various anabolic hormones including sex hormones. *Mimosa hamata* extracts which exhibited antibacterial activity and antiviral activity. It is concluded that both extract could be potential source of active antimicrobial agent.

Table 13.1: Phytochemical Screening of Metanolic Extract of *Mimosa hamata*

Sl.No.	*Tests*	*Leaves of Mimosa hamata*
1.	Tenin	+
2.	Flavonoids	+
3.	Phyto Sterols	+
4.	Protein and amino acids	–
5.	Glycosides	+
6.	Fixed oil and Fat	–
7.	Alkaloid	s +
8.	Quinines	+
9.	Phenols	+
10.	Saponins	+

Conclusion

From above studies, it is concluded that the susceptibility of various microbial agents to different concentrations of *Mimosa hamata* indicates that plant is the potential source for antimicrobial compound. So further work on the profile in order to determine the nature of bioactive principles present in the plant and their mode of action.

In the present era, plant resources are abundant, but these resources are dwindling fast due to the onward march of civilization (Vogel, 1991). Although a significant number of studies have been used to obtain purified plant chemical, very few screening programmes have been initiated on crude plant materials. It has also been widely observed and accepted that the medicinal value of plants lies in the bioactive phytocomponents present in the plants (Veeramuthu *et al.*, 2008).

From the above studies, it is concluded that the traditional plants may represent new sources of anti-microbials with stable, biologically active components that can establish a scientific base for the use of plants in modern medicine. These local ethnomedical preparations and prescriptions of plant sources should be scientifically evaluated and then disseminated properly and the knowledge about the botanical preparation of traditional sources of medicinal plants can be extended for future investigation into the field of pharmacology, phytochemistry, ethnobotany and other biological actions for drug discovery.

References

Ahmad I, and Beg A Z J, 2001. *"Antimicrobial and phytochemical studies on 45 Indian medicinal plants against multi-drug resistant human pathogens." Ethnopharmacol* 74:113-23.

Arthur H R, 1954.*"Phytochemical Analysis of Mimosa* Sps*" J Pharm Pharmacol* 66–72.

Das S, Pal S, Mujib A and Dey S, 1999. *"Biotechnology of medicinal plants- Recent advances and potential."* Ist Edition, Vol II (UK992 Publications, Hyderabad), 126-139.

Deininger R, 1984. *Lectures of the Medical Congress Berlin: Firma Klosterfrau, 31.Germplasm Resources Information Network 2008. "Mimosa pudica L."*. GRIN, United *Koln. 24.*

Dey B and Sita Raman MV, 1957. *"LaboratoryManual of organic Chemistry"*. S. Viswanathan publication, Madras.

Gandhiraja N, Sriram S, Meenaa V, Srilakshmi J K, Sasikumar C and Rajeswari R, 2009. *"Study of different Chemical Components of Mimosa pudica."* Ethnobotanical Leaflets, 13: 618-24.

Krishnaraju AV, Rao TVN, Sundararaju S, Vanisree M, Tsay HS, Gottumukkala V, Subbaraju G, 2006. *"Use of Medicinal plants in Aurveda"* Int. J. Appl. Sci. Eng., 4: 115-125.

Lozoya M, Lozaya X, 1989. *"Weed plants and its uses."n*Tepescohuite Arch. Invest. Mex., pp. 87-93.

Madan Mohan S, Pandey B and Rao Sunita G, 2015. *Phytochemical Analysis and Uses of Mimosa* Sps. *in Chhattisgarh, IOSR Journal of Environmental Science, Toxicology and Food Technology*, Volume. 1 Issue. 3, pp. 01-04.

Ojalla T, Remes S, and Hans P, 1999. *"Antimicrobial activity of some coumarin containing herbal plants growing in Finland". J. Ethnopharmacology*, 68: 267-274.

Oudhia P, 2006. " Preliminary Phytochemical Analysis and Anti-Bacterial Activity of *Mimosa pudica* Linn. Leaves. Available from: Error! Hyperlink reference not valid..edu/newcorp/corpFactSheets/mimosa.html.

Palacious C, and Reyes R E, 1991. *Antibacterial and Antimycotic of Mimosa pudica in experimental animals, Arch Invest Med. (Mex)*, 22(2): 163-169.

Shelef L A, 1983. *Antimicrobial effects of spices*. J. Food Safety 6: 29-U S Forest service. States Deptt. of Agriculture, Agricultural Research Service, Beltsville Area.

Vaidyaratanm P S, 2001. Indian medicinal plants database, 1st edn, Orient Longman, Arya Vidyashala, Kottakkal, II, 36-37.

Zaika L J, 1988. *Evaluation of wound healing activity of root of Mimosa pudica.*

14

Environmental Pollution: Its Effects on Live and its Remedies

Neelima Gupta

Asssistnat Professor, Department of Education, Disha College of Management Studies, Raipur, C.G.

ABSTRACT

Environment pollution is a wide-reaching problem and it is likely to influence the health of human populations is great. This paper provides the insight view about the affects of environment pollution in the perspective of air pollution, water and land/soil waste pollution on human by diseases and problems, animals and trees/plants. Study finds that these kinds of pollutions are not only seriously affecting the human by diseases and problems but also the animals and trees/plants. According to author, still time left in the hands of global institutions, governments and local bodies to use the advance resources to balance the environment for living and initiates the breathed intellectuals to live friendly with environment. As effective reply to contamination is largely base on human appraisal of the problem from every age group and contamination control program evolves as a nationwide fixed cost-sharing effort relying upon voluntary participation (Sharp and Bromley, 1979).

Keywords: *Environment pollution, Air pollution, Water pollution, Soil pollution, Land pollution, Remedies.*

Introduction

The significance of environmental factors to the health and well-being of human populations' is increasingly apparent (Rosenstock 2003; World Health Organization [WHO], 2010b). Environment pollution is a worldwide problem and its potential to influence the health of human populations is great (Fereidoun *et al.*, 2007; Progressive Insurance, 2005.). Pollution reaches its most serious proportions in the densely settled urban-industrial centers of the more developed countries (Kromm, 1973). In poor countries of the world more than 80 per cent polluted water have been used for irrigation with only seventy to eighty percent food and living security in industrial urban and semi urban areas. (Mara and Cairncross, 1989). Industry, clustered in urban and semi-urban areas surrounded by densely populated, low-income localities, continues to pollute the environment with impunity (Government of Pakistan, 2009). Over the last three decades there has been increasing global concern over the public health impacts attributed to environmental pollution (Kimani, 2007), Human exposure to pollution is believed to be more intense now than at any other time in human existence (Schell *et al.*, 2006). Pollution can be made by human activity and by natural forces as well (Fereidoun *et al.*, 2007; The Encyclopedia of the Atmospheric Environment, n.d). Selfish private enterprise and their lack of awareness of public well-being and social costs (Carter, 1985) and natural disasters (Huppart and Sparks, 2006) *e.g.* volcanic ash from Iceland (World Health Organization [WHO], 2010a) are the one of the main reason of pollution. British Airways (1993) expresses their concern about environment in their general goal 'to be a good neighbor, concerned for the community and the environment. This implies that, businesses now adopted this responsibility as part of their overall business strategy; Which should match their broader business goals (Pearce, 1991).

At present, the adoption of environmental auditing in any economic sector is voluntary but future legislation could well make it mandatory (Goodall, 1995). Sharp and Bromley (1979) posit that pollution control program evolves as a Nationwide fixed cost-sharing effort relying upon voluntary participation. Interestingly, Goodall (1995) refers tourism as the potential to damage the environment.

There is no doubt that excessive levels of pollution are causing a lot of damage to human and animal health, plants and trees including tropical rainforests, as well as the wider environment.(Tropical Rainforest Animals, 2008). According to Fereidoun *et al.* (2007), Tehran is one of victim cities in terms of environmental pollution. Gautam *et al.* (2009) nominated Indian cities, among the most polluted cities in the world. Carter (1985) found pollution in formally known Czechoslovakia (now Czech Republic *and* Slovakia) a serious issue which ultimately affects soils and vegetation. As Debarteleven (1992) postulates that environmental pollution and degradation are serious problems in Eastern and Central Europe. Kan (2009) originated the fact about China that, it has environmental problems, including outdoor and indoor air pollution, water shortages and pollution, desertification, and soil pollution, have become more pronounced and are subjecting Chinese residents to significant health risks.

Environmental pollution is tangled with the unsustainable anthropogenic activities, resulting in substantial public health problems. (Khan, 2004). McGeehin *et al.* (2004) reported that U.S. population from infectious diseases to diseases such as cancer, birth defects, and asthma, many of which may be associated with environmental exposures. There is virtually no check on some 8,000 industrial units in USA that are contributing to high rates of pollution (Kaufman, 1993).

Environmental health problems are not simply a conglomerate of concerns about Radiological health, water and wastewater treatment, air pollution control, solid waste disposal, occupational health, etc. (Lynn and Metzler, 1968). The Linton (1967), Spilhaus (1966) and Tukey (1965) made attempt to detail many of the specific environmental health problems which confront contemporary man.

Air Pollution

The air we breathe is an essential ingredient for our wellbeing and a healthy life. Unfortunately polluted air is common throughout the world (EPHA, 2009) especially in developed countries from 1960s. (Kan, 2009). South of Poland (Krzeœlak and Korytkowski, 1994), Ukraine (Avdeev and Korchagin, 1994), China (Kan, 2009), and Pakistan (Government of Pakistan, 2009; Khan, 2010) even famous crowded cities and countries are facing air pollution. Polluted air contains one, or more, hazardous substance, pollutant, or contaminant that creates a hazard to general health (Health and Energy, 2007). The main pollutants found in the air we breathe include, particulate matter, PAHs, lead, ground-level ozone, heavy metals, sulphur dioxide, benzene, carbon monoxide and nitrogen dioxide (European Public Health Alliance, 2009). Air pollution in cities causes a shorter lifespan for city dwellers (Progressive Insurance, 2005). Holland *et al.* (1979) illustrated that British scientists concluded that particulate and related air pollution at high levels pose hazards to Human health. According to Mishra (2003) rapid growth in urban population, increasing industrialization, and rising demands for energy and motor vehicles are the worsening air pollution levels. He added other factors, such as poor environmental regulation, less efficient technology of production, congested roads, and age and poor maintenance of vehicles, also add to the problem. He further added that air pollution is caused of ill health and death by natural and man-made sources, major man-made sources of ambient air pollution include tobacco smoke, combustion of solid fuels for cooking, heating, home cleaning agents, insecticides industries, automobiles, power generation, poor environmental regulation, less efficient technology of production, congested roads, and age and poor maintenance of vehicles. The natural sources include incinerators and waste disposals, forest and agricultural fires (European Public Health Alliance, 2009).

Water Pollution

The water we drink is essential ingredients for our wellbeing and a healthy life. Unfortunately polluted water and air are common throughout the world (European Public Health Alliance, 2009). The WHO states that one sixth of the world's population; approximately 1.1 billion people do not have access to safe water and 2.4 billion lack basic sanitation (European Public Health Alliance, 2009). Polluted water consists of Industrial discharged effluents, sewage water, rain water pollution (Ashraf *et al.*, 2010) and polluted by agriculture or households cause damage to human health or the environment. (European Public Health Alliance, 2009). This water pollution affects the health and quality of soils and vegetation (Carter, 1985). Some water pollution effects are recognized immediately, whereas others don't show up for months or years (Ashraf *et al.*, 2010). Estimation indicates that more than fifty countries of the world with an area of twenty million hectares area are treated with polluted or partially treated polluted water (Hussain *et al.*, 2001) including parts of all continents (Avdeev and Korchagin, 1994; Carter, 1985; Kan, 2009; Khan, 2010; Krzeœlak and Korytkowski, 1994; Wu *et al.*, 1999) and this poor quality water causes Health hazard and death of human being, aquatic life and also disturbs the production of different crops (Ashraf *et al.*, 2010; Scipeeps, 2009). In fact, the effects of water pollution are said to be the leading cause of death for humans across the globe, moreover, water pollution affects our oceans, lakes, rivers, and drinking water, making it a widespread and global concern (Scipeeps, 2009). A drinking water contained a fluoride content ranging from 5.26 to 26.32 milligrams per liter and this is too high as compared to the World Health Organization's standard of 0.6 to 1.7 milligram per liter (Rizvi, 2000). According to Ashraf *et al.* (2010).

In present scenario due to industrialization and increased population, the drains of Pakistan carry the industrial and municipal effluents that are ultimately carried that polluted water to the canals and rivers. The untreated industrial and municipal wastes have created multiple environmental hazards for mankind, irrigation, drinking and sustenance of aquatic life. The drainage water contains heavy metals in addition to biological contaminations. This water pollution infected our food in addition to groundwater contamination when used to irrigate crops.

Pakistani cities are facing tribulations of urban congestion, deteriorating air and water quality and waste management while the rural areas are witnessing rapid deforestation, biodiversity and habitat loss, crop failure, desertification, land degradation, clean drinking water, noise pollution, sanitation (Government of Pakistan, 2009).

Land/Solid Waste Pollution

Improper management of solid waste is one of the main causes of environmental pollution (Kimani, 2007). Land pollution is one of the major forms of environmental catastrophe our world is facing today (Khan, 2004). As Bulgaria and the Slovak Republic, heavy metal industries have produced wastes that are deposited into landfills without special precautions (Lenkova and argova, 1994; Spassov, 1994). Cucu *et al.* (1994) posit that approximately half of the population lives in the vicinity of waste sites that do not conform to contemporary standards in Romania. Czech Republic's coal and uranium mines have produced serious pollution problems, and much of the solid industrial waste containing heavy metals is disposed of, without pretreatment, in open dumps (Rushbrook, 1994). Harvath and Hegedus (1994) concluded as the worst pollution of Hungary comes from open cast mines, lignite-based power plants, chemical factories, and the aluminum industry. The Silesia district in the south of Poland has severe contamination from mining and industry (Krzeœlak and Korytkowski, 1994). Avdeev and Korchagin (1994) conceived soil pollution is critical issues in Ukraine. World Bank (2002) found Particulate matter is the most serious pollutant in large cities in South Asia.

Effects of Dying Environment on Human, Animals and Plants

Environment dying is global perilous point which catastrophically the human, animals and plants. Air pollution results are Cancer (Ries *et al.*, 1999; European Public

Health Alliance, 2009), neurobehavioral disorders (Blaxill 2004; Landrigan *et al.*, 2002; Mendola *et al.*, 2002; Schettler 2002; Stein *et al.*, 2002), cardiovascular problems (European Public Health Alliance, 2009; Tillett, 2009), reduced energy levels (Colls, 2002), premature death (European Public Health Alliance, 2009), asthma (Brauer *et al.*, 2007; Gehring *et al.*, 2002; Jacquemin *et al.*, 2009; Mannino *et al.*, 1998; McConnell *et al.*, 2006; Modig *et al.*, 2006), asthma exacerbations (D'Amato *et al.*, 2005; Heinrich and Wichmann, 2004; Künzli *et al.*, 2000; Nel, 2005;), headaches and dizziness (Colls, 2002), irritation of eyes, nose, mouth and throat (Colls, 2002), reduced lung functioning (Colls, 2002; Gauderman *et al.*, 2005), respiratory symptoms (Colls, 2002; Vichit-Vadakan, 2001), respiratory disease (European Public Health Alliance, 2009; Firkat, 1931), disruption of endocrine (Colls, 2002; Crisp *et al.*, 1998) and reproductive and immune systems (Colls, 2002; European Public Health Alliance, 2009). London Fog episode of 1952, where a sharp increase in particulate matter air pollution led to increased mortality among infants and older adults (Woodruff *et al.*, 2006). High air pollution levels have been linked to infant mortality. (Fereidoun *et al.*, 2007). Air pollutants can also indirectly affect human health through acid rain, by polluting drinking water and entering the food chain, and through global warming and associated climate change and sea level rise. (Mishra, 2003). Associations between particulate air pollution and respiratory disease are reported in Meuse Valley, Belgium, in December 1930 (Firkat, 1931), an episode in Donora, Pennsylvania, in 1948 (Ciocco and Thompson, 1961) and the most notable occurring in December 1952 (Logan, 1953). According to Gardiner (2006) acid rain destroys fish life in lakes and streams and kill trees, destroy the leaves of plants, can permeate soil by making it inappropriate for reasons of nutrition and habitation, unwarranted ultraviolet radiation through the ozone layer eroded by some air pollutants, may cause skin cancer in wildlife and damage to trees and plants, and Ozone in the lower atmosphere may damage lung tissues of animals and can prevent plant respiration by blocking stomata (openings in leaves) and negatively affecting plants' photosynthesis rates which will stunt plant growth; ozone can also decay plant cells directly by entering stomata. Polluted drinking water or water polluted by chemicals produced waterborne diseases like, Giardiasis, Amoebiasis, Hookworm, Ascariasis, Typhoid, Liver and kidney damage, Alzheimer's disease, non-Hodgkin's Lymphoma, multiple Sclerosis, Hormonal problems that can disorder development and reproductive processes, Cancer, heart disease, damage to the nervous system, different type of damages on babies in womb, Parkinson's disease, Damage to the DNA and even death, meanwhile, polluted beach water contaminated people like stomach aches, encephalitis, Hepatitis, diarrhoea, vomiting, gastroenteritis, respiratory infections, ear ache, pink eye and rashes (Water Pollution Effects, 2006). Loss of wild life is directly related to pollution (Progressive Insurance, 2005) and according to Water Pollution Effects (2006) on animals:

1. Nutrient polluted water causes overgrowth of toxic algae eaten by other aquatic animals, and may cause death; it can also cause eruptions of fish diseases.
2. Chemical contamination can cause declines in frog biodiversity and tadpole mass.

3. Oil pollution can increase susceptibility to disease and affect reproductive processes and negatively affect development of marine organisms and it can also a source of gastrointestinal irritation, damage to the nervous system, liver and kidney damage
4. Mercury in water can cause reduced reproduction, slower growth and development, abnormal behavior and death
5. Persistent organic pollutants may cause declines, deformities and death of fish life and Fish from polluted water and vegetable/crops produced or washed from polluted water could also make impact on human and animal health. More sodium chloride (ordinary salt) in water may kill animals and plants, plants may be killed by mud from construction sites as well as bits of wood and leaves, clay and other similar materials and plants may be killed by herbicides in water (Kopaska-Merkel, 2000). For tree and plants water pollution may disrupt photosynthesis in aquatic plants and thus affecting ecosystems that depend on these plants (Forestry Nepal, n.d). Soil pollution effects causes according to tutorvista (n.d) are cancer including leukaemia and it is danger for young children as it can cause developmental damage to the brain furthermore it illustrated that mercury in soil increases the risk of neuromuscular blockage, causes headaches, kidney failure, depression of the central nervous system, eye irritation and skin rash, nausea and fatigue. Soil pollution closely associated to air and water pollution, so its numerous effects come out as similar as caused by water and air contamination. TNAU Agritech Portal (n.d) soil pollution can alter metabolism of plants' metabolism and reduce crop yields and same process with microorganisms and arthropods in a given soil environment; this may obliterate some layers of the key food chain, and thus have a negative effect on predator animal class. Small life forms may consume harmful chemicals which may then be passed up the food chain to larger animals; this may lead to increased mortality rates and even animal extinction.

Conclusion

It appears that polluted environment is global an issue and world community would bear worst results more as they already faced. As effective response to pollution is largely based on human appraisal of the problem (Kromm, 1973) and pollution control program evolves as a nationwide fixed cost-sharing effort relying upon voluntary participation (Sharp and Bromley, 1979). Education, research, and advocacy, are lacking in the region as preventive strategy for pollution (Fitzgerald, 1998) especially in Asia. At present the adoption of environmental auditing in any economic sector is voluntary but future legislation could well make it mandatory (Goodall, 1995) and still time available to use technology and information for environmental health decision. Policymakers in developing countries need to design programs, set standards, and take action to mitigate adverse health effects of air pollution. Healthy people mean human resources are the main object of any successful business or country. These societal beneficial efforts need to carefully adapt available knowledge from other settings, keeping in mind the differences in pollutant mixtures, concentration levels, exposure patterns, and various underlying population characteristics.

References

Ashraf, M. A., Maah, M. J., Yusoff, I. and Mehmood, K. (2010). Effects of Polluted Water Irrigation on Environment and Health of People in Jamber, District Kasur, Pakistan, *International Journal of Basic and Applied Sciences,* 10(3), pp. 37-57.

Avdeev, O. and Korchagin, P. (1994). Organization and **Implementation** of Contaminated Waste Neutralization in the Ukraine - National Report II, *Central. European Journal* of *Public Health,* 2(suppl), pp. 51-52.

Blaxill, M. F. (2004). What's going on? The Question of Time Trends in Autism. *Public Health Reports,* 119(6), pp. 536-551.

Brauer, M., Hoek, G., Smith, H. A., de Jongste, J. C., Gerritsen, J. and Postma, D. S. (2007). Air Pollution and Development of Asthma, Allergy and Infections in a Birth Cohort, *European Society for Clinical Respiratory Physiology, 29*(5), pp. 879-888.

British Airways, (1993). *Annual Environment Report,* London: British Airways plc, Environment Branch, Heathrow.

Carter, F. W. (1985). Pollution Problems in Post-War Czechoslovakia, *Transactions of the Institute of British Geographers,* 10(1), pp. 17-44.

Ciocco, A. and Thompson, D. J. (1961). A Follow-up on Donora Ten Years After: Methodology and Findings, *American Journal of Public Health,* 51(2), pp. 155- 164.

Colls, J. (2002). *Air Pollution.* New York: Spon Press.

Crisp, T. M., Clegg, E. D., Cooper, R. L., Wood, W. P., Anderson, D. G., Baetcke, K. P., Hoffmann, J. L., Morrow, M. S., Rodier, D. J., Schaeffer, J. E., Touart, L. E., Zeeman, M. G. and Patel, Y. M. (1998). Environmental Endocrine Disruption: An Effects Assessment and Analysis, *Environmental Health Perspectives,*106(1), pp. 11-56.

Cucu. M., Lupeanu, M. I., Nicorici, M., Lonescu, L. and Sandu, S. (1994). The Dangerous Wastes and Health Risks in Romania: National Report, *Central European Journal of Public Health,* 2(suppl), pp. 41-43.

D'Amato, G., Liccardi, G., D'Amato, M. and Holgate. S. (2005). Environmental Risk Factors and Allergic Bronchial Asthma, *Clinical and Experimental Allergy,* 35(9), pp.1113- 1124.

DeBarteleven, J. (1992). *Eastern Europe's Environmental Crisis.* Baltimore, MD: Johns Hopkins Press.

European Public Health Alliance, (2009). *Air, Water Pollution and Health Effects.* Retrieved from http://www.epha.org/r/54

Fereidoun, H., Nourddin, M. S., Rreza, N. A., Mohsen, A., Ahmad, R. and Pouria, H. (2007). The Effect of Long-Term Exposure to Particulate Pollution on the Lung Function of Teheranian and Zanjanian Students, *Pakistan Journal of Physiology*, 3(2), pp. 1-5.

Firket, J. (1931). The Cause of the Symptoms Found in the Meuse Valley during the Fog of December, *Bulletin de l'scademie Royale Medicine de Belgique*, 11, pp. 683-741.

Fitzgerald, E. F., Schell, L. M., Marshall, E. G., Carpenter, D. O., Suk. W. A. and Zejda, J. E. (1998). Environmental Pollution and Child Health in Central and Eastern Europe, *Environmental Health Perspectives*, 106(6), pp. 307-311.

Forestry Nepal (n.d) *Pollution Effects on Plants and Trees*, Retrieved from http://www.forestrynepal.org/notes/silviculture/locality-factors/16

Gardiner, L. (2006). *Air Pollution Affects Plants, Animals, and Environments*. Windows to the Universe. Retrieved from http://www.windows.ucar.edu/tour/link=/earth/Atmosphere/wildlife_forests.html and edu=high

Gauderman, W. J., Avol, E., Gilliland, F., Vora, H., Thomas, D., Berhane, K., McConnell. R., Kuenzli, N., Lurmann, F., Rappaport, E., Margolis, H., Bates, D. and Peters, J. (2005). The Effect of Air Pollution on Lung Development from 10 to 18 Years of Age, *New England Journal of Medicine*, 352(12), pp. 1276.

Gautam, A., Mahajan, M. and Garg, S. (2009). *Impact of Air Pollution on Human Health in Dehra Doon City*, Retrieved from http://www.esocialsciences.com/data/articles/Document12882009311.130313E- 02.pdf

Gehring, U., Cyrys, J., Sedlmeir, G., Brunekreef, B., Bellander, T, and Fischer, P. (2002). Traffic Related Air Pollution and Respiratory Health During the First 2 Years of Life. *European Respiratory Journal*, 19(4), pp. 690-698.

Goodall. B. (1995). Environmental Auditing: A Tool for Assessing the Environmental Performance of Tourism Firms, *The Geographical Journal*, 161(1), pp. 29-37.

Government of Pakistan (2009). *Economic Survey of Pakistan*, Finance Division, Economic Division Wing, Islamabad.

Harvath, A. and Hegedus, E. (1994). Hazardous Wastes in Hungary-National Report, *Central European Journal of Public Health*, 2(suppl), pp. 30-33.

Health and Energy, (2007). *Air Pollution Health Effects*, Retrieved from http://healthandenergy.com/air_pollution_health_effects.htm

Heinrich, J. and Wichmann, H. E. (2004). Traffic Related Pollutants in Europe and Their Effect on Allergic Disease, *Current Opinion* in *Allergy* and *Clinical Immunology*, 4(5), pp. 341-348.

Holland, W. W., Bennett, A. E., Cameron, I. R., Florey, C. V., Leeder, S. R., Shilling, R. S. F., Swan, A. V. and Waller, R. E. (1979). Health Effects of Particulate Pollution: Reappraising the Evidence. *Am Journal Epidemiol*, 110(5), pp. 525- 659.

Huppert, H. E. and Sparks, R. S. J. (2006). Extreme Natural Hazards: Population Growth, Globalisation and Environmental Change, *Philosophical. Transactions of the Royal Society*, 364(1845), pp. 1875-1888.

Hussain, I., Raschid, L., Hanjra, M. A., Marikar, F. and van der Hoek, W. (2001). *A Framework* for Analyzing Socioeconomic, Health and Environmental Impacts of Wastewater Use in *Agriculture in Developing Countries*, IWMI

Jacquemin, B., Sunyer, J., Forsberg, B., Aguilera, I., Briggs, D. and Garcia-Esteban, R. (2009). Home Outdoor $N0_2$ and New Onset of Self Reported Asthma in Adults. *Epidemiology*, 20(1), pp. 119-126.

Kan, H. (2009). Environment and Health in China: Challenges and Opportunities *Environmental Health Perspectives*, 117(12), pp. A530-A531

Kaufman, B. E. (1993). *The Origins and Evolution of the Field of Industrial Relations in the United States*, Ithaca, NY, ILR Press.

Khan, A. (2010). *Air pollution in Lahore*, The Dawn, Retrieved from http://news.dawn.com/wps/wcm/connect/dawn-content-library/dawn/the-newspaper/letters-to the-editor/air-pollution-in-lahore-070

Khan, S. I. (2004). *Dumping of Solid Waste: A Threat to Environment*, The Dawn, Retrieved from http://66.219.30.210/weekly/science/archive/040214/science13.htm

Kimani, N. G. (2007). *Environmental Pollution and Impacts on Public Health: Implications of the Dandora Dumping Site Municipal in Nairobi, Kenya*, United Nations Environment Programme, pp. 1-31. Retrieved from http://www.korogocho.org/english/index.php?option=com_docman and task=doc_download and gid=54 and Itemid=73

Kopaska-Merkel, D. (2000). *How Does Water Pollution Affect Plant Growth?* Mad Sci Network. Retrieved from http://www.madsci.org/posts/archives/2000-11/974847556.En.r.html

Kromm, D. E. (1973). Response to Air Pollution in Ljubljana, Yugoslavia, *Annals of the Association of American Geographers*, 63(2), pp. 208-217.

Krzeœlak, A. and Korytkowski, J. (1994). Hazardous Wastes in Poland-National Report, *Central European Journal of Public Health*, 2(suppl), pp. 44-40.

Künzli N., Kaiser, R., Medina, S., Studnicka, M., Chanel, O., Filliger, P., Herry, M., Horak, F. Jr., Puybonnieux-Texier, V., Quénel, P., Schneider, J., Seethaler, R., Vergnaud, J-C. and Sommer, H. (2000). Public Health Impact of Outdoor and Traffic Related Air Pollution: A European Assessment, *The Lancet*, 356(9232), pp. 795-801.

Landrigan, P. J., Schechter, C. B., Lipton, J. M., Fahs, M. C. and Schwartz, J. (2002). Environmental Pollutants and disease in American Children: Estimates of Morbidity, Mortality, and Costs for Lead Poisoning, Asthma, Cancer, and Developmental disabilities, *Environmental Health Perspectives*, 110(7), pp. 721-728.

Lenkova, K. and Vargova, M. (1994). Hazardous Wastes in the Slovak Republic-National Report, *Central European Journal of Public Health*, 2(suppl), pp. 43-48.

Linton, R. M. (1967). *A Strategy for a Livable Environment, Report of the Secretary's Task Force on Environmental Health and Related Problems*, U. S. Department of Health, Education, and Welfare, Washington, DC.

Logan, W. P. D. and Glasg, M. D. (1953). Mortality in London Fog Incident, 1952, *The Lancet*, 261(6755), pp. 336-338.

Lynn, W. R. and Metzler, D. F. (1968). Environmental Health Decision-Making, *Journal-Water Pollution Control Federation*, 40(7), pp. 1311-1313.

Mannino, D. M., Homa, D. M., Pertowski, C. A., Ashizawa, A., Nixon, L. L., Johnson, C. A.,Ball, L. B., Jack, E. and Kang, D. S. (1998). Surveillance for Asthma-United States, 1960-1995. *MMWR CDC Surveillance Summaries*, 47(1), pp.1-27.

Mara, D. and Cairncross, S. (1989). *Guidelines for Safe Use of Wastewater and Excreta in Agriculture and Aquaculture: Measures for Public Health Protection*. World Health Organization, Geneva, pp.187.

McConnell, R., Berhane, K., Yao, L., Jerrett, M., Lurmann, F. and Gilliland, F. (2006). Traffic, Susceptibility, and Childhood Asthma. *Environmental Health Perspective*, 114(5), pp. 766-772.

McGeehin, M. A., Qualters, J. R. and Niskar, A. S. (2004). National Environmental Public Health Tracking Program: Bridging the Information Gap, *Environmental Health Perspectives*, 112(14), pp. 1409-1413

Mendola, P., Selevan, S. G., GUttar, S. and Rice, D. (2002). Environmental Factors Associated with a Spectrum of Neurodevelopmental Deficits. *Mental Retardation and Developmental Disabilities Research Reviews*, 8(3), pp. 188-197.

Mishra, V. (2003). Health Effects of Air Pollution, Background paper for Population-Environment Research Network (PERN) Cyberseminar, December 1-15. Retrieved from http://www.mnforsustain.org/climate_health_effects_of_air_pollution_mishra_pern.htm

Modig, L., Jarvholm, B., Ronnmark, E., Nystrom, L., Lundback, B. Andersson, C. and Forsberg, B. (2006). Vehicle Exhaust Exposure in an Incident Case Control Study of Adult Asthma. *European Respiratory Journal*, 28(1), pp. 75-81

Nel, A. (2005). Air pollution Related Illness: Effects of Particles. *Science*, 308(5723), pp. 804-806.

Pearce, D. (1991). *Corporate Responsibility and the Environment*. London: British Gas.

Progressive Insurance, (2005). Pollution Impact on Human Health. Retrieved from http://www.progressiveic.com/n25feb05.htm

Rizvi, M. (2000). *Bone Disease Spurs Pakistan to Environmental Action*, Fluoride Action Network, Retrieved from http://www2.fluoridealert.org/Alert/Pakistan/Bone-disease-spurs-Pakistan-to- environmental-action

Ries, L. A. G., Smith, M. A., Gurney, J. G., Linet, M., Tamra, T., Young, J. L. and Bunin, G. R. (eds) (1999). *Cancer Incidence and Survival among Children and*

Adolescents: United States SEER Program 1975-1995, Bethesda, MD, National Cancer Institute, SEER Program.

Rosenstock, L. (2003). The Environment as a Cornerstone of Public Health, *Environmental Health Perspectives*, 111(7), pp. A376-A377.

Rushbrook, P. (1994). Regional Health Issues Related to Hazardous Wastes, *Central European Journal of Public Health*, 2(suppl), pp. 16-20.

Schell, L. M., Gallo, M. V., Denham, M., and Ravenscroft, J. (2006). Effects of Pollution on Human Growth and Development: An Introduction, *Journal of Physiological Anthropology*, 25(1), 103-112.

Schettler, T. (2002). Changing Patterns of Disease: Human Health and the Environment,.*San Francisco Medicine*, Retrieved from http://www.sfms.org/AM/Template.cfm?Section=Home and SECTION=Article_Archives and CONTENTID=1643 and TEMPLATE=/CM/HTMLDisplay.cfm

Scipeeps, (2009). Effects of Water Pollution. Retrieved from http://scipeeps.com/effects-ofwater-pollution/

Sharp, B. M. H. and Bromley, D. W. (1979). Agricultural Pollution: The Economics of Coordination, *American Journal of Agricultural Economics* 61(4), pp. 591-600.

Spassov, A. (1994). Identification of Problem Related to Solid Wastes in Bulgaria-National Report. *Central European Journal of Public Health*, 2(suppl), pp. 21-23.

Spilhaus, A. (1966). *Waste Management and Control*, Report of the Committee on Pollution, National Academy of Sciences-National Research Council, Washington, D. C.

Stein, J., Schettler, T., Wallinga, D. and Valenti, M. (2002). In Harm's Way: Toxic Threats to Child Development, *Journal* of Developmental and *Behavioral Pediatrics*, 23(0), pp. S13-S22.

The Encyclopedia of the Atmospheric Environment, (n.d). *Impacts of Air Pollution*. Retrieved from http://www.ace.mmu.ac.uk/eae/air_quality/Younger/Impacts.html

TNAU Agritech Portal. (n.d) *Soil Pollution*, TamilNadu Agricultural University, Coimbatore. Retrieved from http://agritech.tnau.ac.in/environment/envi_pollution_intro per cent 20- per cent 20soil.html

Tropical Rainforest Animals, (2008). Pollution Effects on Humans, Animals, Plants and the Environment. Retrieved from http://www.tropical-rainforest-animals.com/pollution-effects.html

Tukey, J. W., Alexander, M., Bennett, H. S., Brady, N. C., Calhoun, J. C. Jr., Geyer, J. C., Haagen-Smit, A. J., Hackerman, N., Hartgering, J. B., Pimentel, D., Revelle, R., Roddis, L. H., Stewart, W. H. and Whittenberger, J. L. (1965). *Restoring the Quality of Our Environment*, Report of the Environmental Pollution Panel, President's Science Advisory Committee, Washington, D.C.

Tutorvista, (n.d). *Consequences of Soil Pollution*, Retrieved from http://www.tutorvista.com/english/consequences-of-soil-pollution

Vichit-Vadakan, N., Ostro, B. D., Chestnut, L. G., Mills, D. M., Aekplakorn, W., Wangwongwatana, S. and Panich, N. (2001). Air Pollution and Respiratory Symptoms: Results from Three Panel Studies in Bangkok, Thailand, *Environmental Health Perspectives,* 109(3), pp. 381-387

Water Pollution Effects, (2006). In *Grinning Planet, Saving the Planet One Joke at a Time*. Retrieved from http://www.grinningplanet.com/2006/12-05/water-pollution-effects.htm

Woodruff, T. J., Parker, J. D. and Schoendorf, K. C. (2006). Fine Particulate Matter (PM2.5) Air Pollution and Selected Causes of Post neonatal Infant Mortality in California, *Environmental Health Perspectives*. 114(5), pp. 786–790.

World Bank, (2002). What Do We Know About Air Pollution?—India Case Study, Urban Air Pollution, South Asia Urban Air Quality Management Briefing Note No. 4, pp. 1-4.

World Health Organization (WHO), (2010a). Air Quality: Volcanic Ash Cloud over Europe. Retrieved from http://www.euro.who.int/en/what-we-do/health-topics/environmental-health/airquality/volcanic-ash-cloud-over-europe

World Health Organization (WHO), (2010b). The World Health Report - Health Systems Financing: The Path to Universal Coverage. Retrieved from

http://www.who.int/entity/whr/2010/whr10_en.pdf

Wu, C., Maurer, C., Wang, Y., Xue, S. and Davis, D. L. (1999). Water Pollution and Human Health in China, *Environmental Health Perspectives,* 107(4), pp. 251-256.

15

Solid Waste Management

Manisha Garg

Arts and Commerce Girls College,
Devendra Nagar, Raipur, C.G.

Introduction

Solid waste is the unwanted or useless solid material generated from combined residential, industrial and commercial activities in a given area. It may be categorized according to its origin (domestic, industrial, commercial, construction or institutional); according to its contents (organic material, glass, metal, plastic paper etc); or according to hazard potential (toxic, non-toxin, flammable, radioactive, infectious etc).

Reduce, Reuse, Recycle

Methods of waste reduction, waste reuse and recycling are the preferred options when managing waste. There are many environmental benefits that can be derived from the use of these methods. They reduce or prevent green house gas emissions, reduce the release of pollutants, conserve resources, save energy and reduce the demand for waste treatment technology and landfill space. Therefore it is advisable that these methods be adopted and incorporated as part of the waste management plan.

Waste Reduction

Solid waste management is a term for garbage management. As long as humans have been living in settled communities, solid waste, or garbage, has been an issue, and modern societies generate far more solid waste than early humans ever did. Daily life in industrialized nations can generate several pounds (kilograms) of solid waste per consumer, not only directly in the home, but indirectly in factories that manufacture goods purchased by consumers. Solid waste management is a system

Best
option
Reduce waste
Re-use
Recycle / Compost
Incineration with
energy recovery
Incineration
Landfill
with energy recovery
Landfill
Worst option

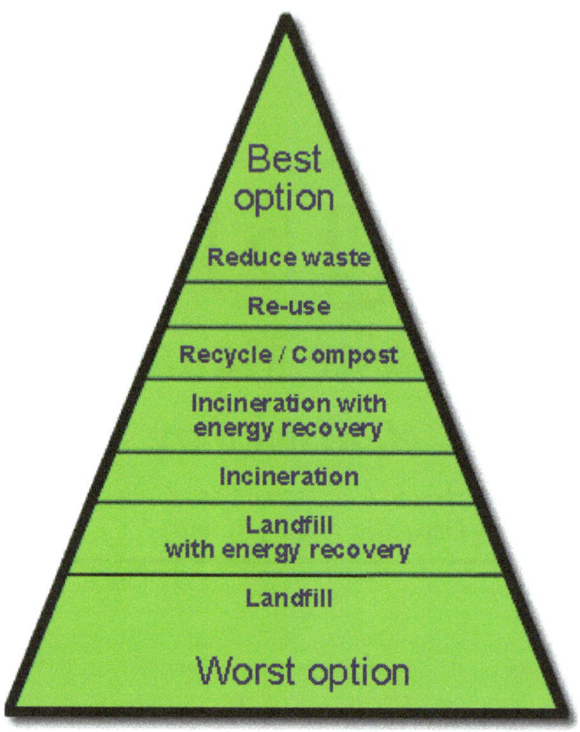

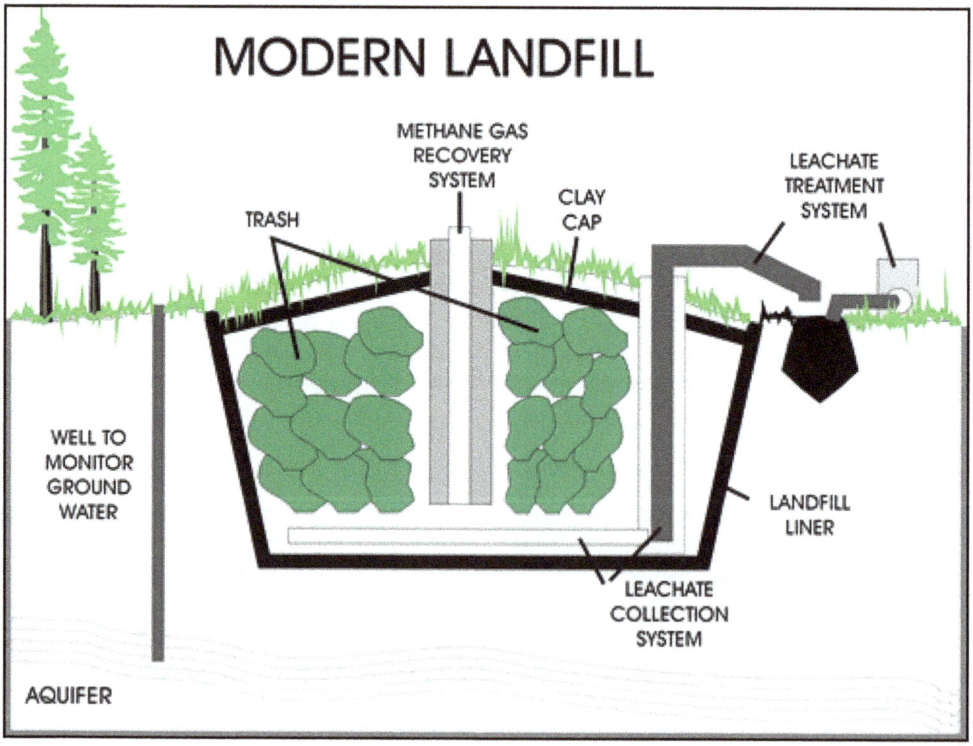

MODERN LANDFILL
METHANE GAS
RECOVERY
SYSTEM
CLAY
CAP
TRASH
LEACHATE
TREATMENT
SYSTEM
WELL TO
MONITOR
GROUND
WATER
LANDFILL
LINER
LEACHATE
COLLECTION
SYSTEM
AQUIFER

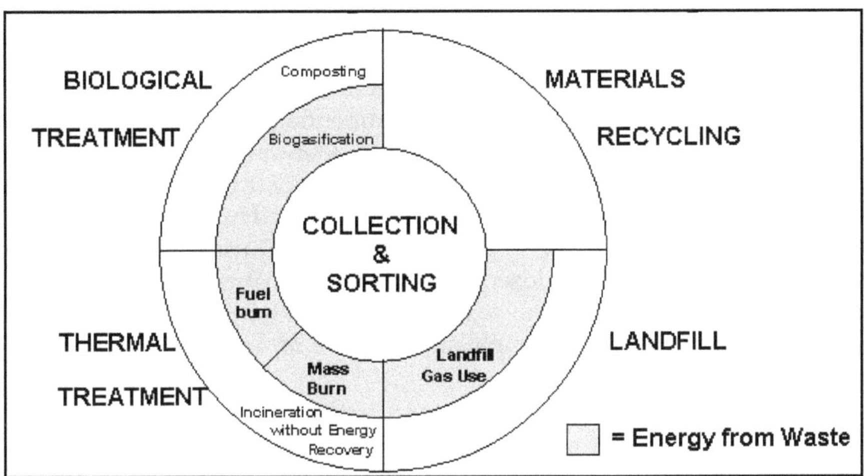

for handling all of this garbage, and includes municipal waste collection, recycling programs, dumps, and incinerators.

To the great benefit of archeology, early methods of waste management consisted of digging pits and throwing garbage into them. This created a record of the kinds of lives that people lived, showing things like what people ate, the materials used to make eating utensils, and other interesting glimpses into historic daily life. When human cities began to be more concentrated, however, dealing with the garbage became a serious issue. Houses that did not have room to bury their trash would throw it into the streets, making a stroll to the corner store an unpleasant prospect. In response, many cities started to set up municipal garbage collection, in the form of rag and bone men who would buy useful garbage from people and recycle it, or waste collection teams that would dispose of unusable waste.

Conclusion and Suggestion

In summary, there are four main ways that most city governments in developing countries can enhance waste reduction:

- **Inform citizens about source separation and recycling, and the needs of waste workers:** extensive public education is needed to develop understanding of the need for further source separation to improve the potential for composting and to remove the stigma of association with waste materials.
- **Promote recycling industries and enterprises.**
- **Divert organics.** The greatest relief for the waste authority will come from reduction of organics, which implies, in the main, successful composting. Keeping organics pure for composting will require more thorough source separation than is done at present.

☆ **Advocate key areas for waste reduction at the manufacturing level** (*e.g.*, reduction of plastic packaging; coding of plastics to improve recycling).

Waste generation and waste reduction reflect many complex economic and social factors. No city or town can adopt recommendations in a vacuum; each must examine its own wastes, and the potential for extending waste reduction. There are many possible ways to implement the general dictum that waste reduction should be the first principle of solid waste management. Humane concern for waste workers must temper the drive to greater efficiency. During periods of technical change, there are winners and losers, and in the field of materials recovery there should be atten- tion to those who lose outÓ as operations become more efficient. In most cases, the resulting municipal strategy will be a mix of private and public sector activities.

References

http://www.solid.gov.bb/Resources/Brochures/Programme/program02.asp >

http://www.wasteonline.org.uk/resources/InformationSheets/WasteDisposal.htm

http://www.eia.doe.gov/kids/energyfacts/saving/recycling/solidwaste/landfiller.html

http://web.mit.edu/urbanupgrading/upgrading/issues-tools/issues/wastecollection. html#Anchor-Collection-45656

http://www.sustainability-ed.org/pages/example2-2.htm

http://www.massbalance.org/downloads/projectfiles/1826-00237.pdf

http://msw.cecs.ucf.edu/Lesson8-Incineration.html

http://www.csiro.au/org/ps16w.html

http://viso.ei.jrc.it/iwmlca/

Sewerage and Solid Waste Project Unit. 2000. *The solid waste management programme.* Sewerage and Solid Waste Project Unit, Barbados.

United Nations Environment Programme *International Source Book on Environmentally Sound Technologies (ESTs) for Municipal Solid Waste Management (MSWM)* <http://www.unep.or.jp/ietc/ESTdir/Pub/MSW/index.asp

16

Sustainable Development in India with Reference to Agricultural Area

Bhawana Thakur

Department of Education,
Disha College of Management Studies, Raipur, C.G.

ABSTRACT

India has been witnessing a blinding pace of growth and development in recent times. There is talk of the country leapfrogging into the league of developed nations sooner than later. But this growth has raised concerns from sundry quarters as regards its basic texture and health. Experts are now calling for "sustainable development" and the term has gained currency in the last few years. In spite of fast growth in various sectors, agriculture remains the backbone of the Indian economy. This paper attempts to tackle and explore the issue of sustainable development in agriculture in India. Further it aims to compare the sustainable agriculture system with the traditional system and the current system in practice, across the dimensions of ecological, economic and social sustainability.It also tries to give long term solutions to solve the problems plaguing the system so that sustainable practices can be promoted and practiced.

Keywords: *Sustainable development, Agriculture, Ecological sustainability, Economic sustainability, Social sustainability.*

Introduction

Agriculture occupies the most important position in Indian economy, as it is one of the largest private enterprises in India, which continues to dominate the change in economy through its links of various sectors of production and markets. The role of agricultural sector in Indian economy can be seen through its contribution to GDP (Gross Domestic Product) and employment. This sector also contributes significantly to sustainable economic development of the country. The sustainable agriculture development of any country depends upon the judicious mix of their available natural resources. In fact agriculture determine the fate of a country like India where about two-thirds of the population still lives in rural India with agriculture as its livelihood, in spite of the increasing urbanization that has been taking place since many decades. Therefore if agriculture goes wrong, it will be really bad for the economy as the falling of agricultural growth not only affects employment but GDP too (thus increasing poverty). The larger objective for the improvement of agriculture sector can be realized through rapid growth of agriculture which depends upon increasing the area of cultivation, cropping intensity and productivity. But for a country like India, increasing productivity is more important than the rest of the two. This is simply because of increasing urbanization, industrialization and the limited land size of the country.

The productivity can be increased by two ways. First, increasing output by efficient utilization of available resources. Second, increasing output by variation of input. The first method is better with respect to productivity and sustainability. But due to increasing population, this method can not provide a permanent solution. Thus we can go for the second method which may potentially cause environmental degradation in the economy and affect its sustainability. Therefore there is need to tackle the issues related to sustainable agriculture development.

Sustainable Agriculture Development

The issues of sustainable development can be discussed under three broad types of farming systems *viz.* traditional production system, modern agriculture system and sustainable agriculture system. Further we can compare them across three dimensions, ecological, economic and social sustainability.

Ecological Sustainability

Most of the traditional and conventional farm practices are not ecologically sustainable. They misuse natural resources, reducing soil fertility causing soil erosion and contributing to global climatic change. But sustainable agriculture has some major advantages over traditional practices:

Soil Fertility

Continuous fall in soil fertility is one of the major problems in many parts of India. Sustainable agriculture improves fertility and soil structure.

Water

Irrigation is the biggest consumer of fresh water, and fertilizer and pesticides contaminate both surface and ground water. Sustainable agriculture increase the organic matter content of the top soil, thus raising its ability to retain and store water that falls as rain.

Biodiversity

Sustainable agriculture practices involve mixed cropping, thus increasing the diversity of crops produced and raising the diversity of insects and other animals and plants in and around the fields.

Health and Pollution

Chemicals, pesticides and fertilizers badly affect the local ecology as well as the population. Indiscriminate use of pesticides, improper storage etc. may lead to health problems. Sustainable agriculture reduces the use of hazardous chemical and control pests.

Land use Pattern

Over-exploitation of land causes erosion, land slides and flooding clogs irrigation channels and reduces the arability of the land. Sustainable agriculture avoids these problems by improving productivity, conserving the soil etc.

Climate

Conventional agriculture contributes to the production of green house gases in various ways like reducing the amount of carbon stored in the soil and in vegetation, through the production of Methane in irrigated field and production of artificial fertilizers etc. By adopting sustainable agriculture system, one can easily overcome this problem.

Economic Sustainability

For agriculture to be sustainable it should be economically viable over the long term. Conventional agriculture involves more economic risk than sustainable agriculture in the long term. Sometimes governments are inclined to view export-oriented production systems as more important than supply domestic demands. This is not right. Focusing on exports alone involves hidden costs: in transport, in assuring local food security, etc. Policies should treat domestic demand and in particular food security as equally important to the visible trade balance.

It is a popular misconception that specific commodities promise high economic returns. But market production implies certain risks as markets are fickle and change quickly. Cheap foreign food may sweep into the national market, leaving Indian farmers without a market. As a World Trade Organization signatory, the Indian government is under pressure to deregulate and open its economy to the world market so it cannot protect its farmers behind tariff walls.

The main source of employment for rural people is farming. Trends towards specialization and mechanization may increase narrowly measured "efficiency", but they reduce employment on the land. The welfare costs of unemployment must be taken into account when designing national agricultural support programs. Sustainable agriculture, with its emphasis on small-scale, labor-intensive activities, helps overcome these problems.

Social Sustainability

Social sustainability in farming techniques is related to the ideas of social acceptability and justice. Development can not be sustainable unless it reduces poverty. The government must find ways to enable the rural poor to benefit from agriculture development. Social injustice is where some section of the society is neglected from development opportunities. But having robust system of social sustainability can bridge the gap between "haves" and „have-nots". Many new technologies fail to become applicable in agriculture sector due to lack of acceptability by the local society. Sustainable agriculture practices are useful because it is based on local social customs, traditions and norms etc. Because of being familiar the local people are more likely to accept and adopt them.Moreover, sustainable agriculture practices are based on traditional know-how and local innovation. Local people have the knowledge about their environment crops and livestock.

Traditional agriculture is more gender oriented, where woman bear the heaviest burden in terms of labor. Sustainable agriculture ensures that the burden and benefits are shared equitably etween man and woman. While conventional farming focuses on a few commodities, sustainable agriculture improves food security by improving quality and nutritional value of food, and also by producing bigger range of products throughout the years. Traditional farming was also driven by the caste and wealth oriented people. The rich and higher castes benefitted more, while the poor and lower castes are left out. Sustainable agriculture attempts to ensure equal participation which recognizes the voice and speech of every people.

Indian Agriculture Sector

Agriculture is one of the most preeminent sectors of the Indian economy. It is the source of livelihood for almost two third of the rural population workforce in the country residing in rural areas. Indian agriculture provides employment to 65 per cent of the labor force, accounts for about 27 per cent of GDP, contributes 21 per cent of total exports and raw material to several industries. The livestock sector contributes an estimated 8.4 per cent to the country GDP and 35.85 per cent of the agriculture output. In India about 75 per cent people are living in rural areas and are still dependent on agriculture, about 43 per cent of India's geographical area is used for agriculture activities. The estimated food grain production is about 211.17 metric tons in the country.

Table 16.1: India's Position in World's Agriculture

Total Area	7th
Irrigated Area	1st
Population	2nd
Economically active population	2nd
Total Cereals	3rd
Wheat production	2nd
Rice Production	2nd
Milk	1st
Livestock (Buffaloes, Castles)	1st
Fish	7th
Production of Inland Fish	2nd

Source: NIC.

The total geographical area comes under the agriculture are 329 MH out of which 265MH represent varying degree of potential production. The net sown area is 143 MH out of which 56MH are net irrigated area in the country.

India is a vast country with variety of land forms, climate, geology, physiography and vegetation. India is endowed with regional diversities for its uneven economic and agriculture development on account of:

- Agro-Climate Environment
- Agro-Ecological Regions
- Agro-Edaphic regions
- Natural Tesource Development
- Human Resource Development
- Level of Investment
- Technological Development.

Agricultural Production in India

Indian Agriculture production in most part of the country is closely related to the optimum use of available natural and human resources of the country. Therefore riding on the back of agro climatic condition and rich natural resource base, India today has become the world's largest producer of numerous commodities. The country is a leading producer of coconuts, mangoes, milk, bananas, dairy products, ginger, turmeric, cashew nut, pulses and black pepper. It is also the second largest producer of rice, wheat, sugar, cotton, fruit and vegetables.

Indian agriculture production is closely related to sufficient and wise water management practices. Most of the agriculture practices in India confined to a few monsoon months. During the monsoon season, India is usually endowed with generous rainfall; although not infrequently, this bountiful monsoon turns into terror, causing uncontrollable floods in different parts of the country and ultimately affecting agriculture production.

Mile Stones in Indian Agriculture

Policy makers and planners, concerned about national independence, security and political stability realized that self sufficiency in food production was an absolute pre requisite for sustainable agriculture development. The policies considered to be a mile stone in agriculture development of the country are:

1. **Green Revolution (1968):** This revolution includes packages of programs like, Intensive Agriculture District Program (IADP) which eventually led to the Green Revolution. The National Bank for Agriculture Development (NABARD) was set up. The emphasis was on high yielding varieties along with other modern inputs like chemicals, fertilizers, pesticides and mechanization and also on how productivity could be raised in agriculture sector without having substantial influences on increasing area under cultivation.

2. **Ever Green Revolution (1996):** Father of Indias Green revolution, Prof. M.S. Swaminathan claims to be pro-woman, pro-nature and pro-poor. The conservation of biodiversity, maintaining soil fertility, increasing the climate resistance of food crops combined with better and more education and technological innovation are the key to the ever green revolution. The main aim of this revolution is to produce more using less land, less water and less fertilizer. The recent visit of US President in New Delhi in March 2010, announced a new partnership with India in an agriculture sector for an evergreen revolution to achieve global food security.
3. **White and Yellow Revolution:** The Green Revolution generated a mood of self confidence in our agriculture capability, which led to the next phase characterized by the Technology Mission. Under this approach, the focus was on conservation, consumption, and commerce. An end-to-end approach was introduced involving attention to all links in the production-consumption chain, owing to which progress was steady and sometimes striking as in the case of milk and egg production.
4. **Blue Revolution (Water, Fish):** It has been brought about in part by a trend towards healthier eating which has increased the consumption of Fish. Additionally the supply of wild fish is declining. This revolution could give landless laborers and women a great opportunity for employment which empowered them.
5. **Bio-Technology Revolution:** India is well positioned to emerge as a significant player in the Global Bio-tech Arena. Agriculture biotech in India has immense growth opportunity and the country could become the fore runner in the transgenic production rise and several other genetically engineered vegetables by 2010.In agri-biotech sector India has been growing at a blinding rate of 30 per cent since the last five years. The food processing sectors which is considered to be prime drivers of Indian economy is currently growing at 13.5 per cent.

Impact of Economic Reform on Indian Agriculture

The Indian agriculture sector has been undergoing economic reform since 1990s in a move to liberalize the economy to benefit from globalization. India, which is one of the largest agriculture based economies, remained closed until the early 1990s. In 1991, the new economic policies stressed both external sector reforms in the exchange rate, trade and foreign investment policies and internal reform in areas such as industrial policies, price and distribution controls, and fiscal restructuring in the financial and public sector.

India's economic reforms were initiated in June 1991, but it was observed that the expected increase in exports due to liberalization did not occur. In addition, the agriculture sector's output growth decreased during 1992-1993 to 1998-1999. The reason behind this was the decline in the environmental quality of land which reduced the marginal productivity of the modern inputs. Agriculture sector is the mainstay of the Indian economy around which socio-economic privileges and deprivation revolve, and any change in its structure is likely to have a corresponding impact on the existing pattern of social equality. No strategy of economic reform can succeed without sustained and broad based agriculture development, which is critical for raising living standards, alleviating poverty, assuring food security, making substantial contribution to the national economic growth. Since agriculture continues to be a tradable sector, this economic liberalization and reform policy has a far reaching effect on:

- ☆ Agricultural exports and imports
- ☆ Investment in new technologies
- ☆ Pattern of agricultural growth
- ☆ Agricultural income and employment
- ☆ Agricultural price
- ☆ Food security

Reduction in Commercial Bank credit to agriculture, in lieu of this reforms process and recommendations of Khusro Committee and Narasimham Committee resulted in fall in farm investment and impaired growth.

Liberalization of agriculture and open market operations enhance competition in "resource use" and "marketing of agriculture production", which forces the small and marginal farmers to resort to " distress sale" and seek off farm employment for supplementing income.

Issues and Challenges

The central issue in agricultural development is the necessity to improve productivity, generate employment and provide a source of income to the poor segments of population. Studies by FAO have shown that small farms in developing countries contribute around 30-35 per cent to the total agricultural output. The pace of adoption of modern technology in India is slow and the farming practices are too haphazard and unscientific. Some of the basic issues for development of Indian agriculture sector are revitalization of cooperative institutions, improving rural credits, research, human resource development, trade and export promotion, land reforms and education.

Future Prospects and Solution for India

Agriculture sector is an important contributor to the Indian economy around which socio-economic privileges and deprivations revolve and any change in its structure is likely to have a corresponding impact on the existing pattern of social equity. Sustainable agricultural production depends upon the efficient use of soil, water, livestock, plant genetics, forest, climate, rainfall and topology. Indian agriculture faces resource constraints, infrastructural constraints, institutional constraints, technological constraints and policy induced limitations.

Sustainable development is the management and conservation of the natural resource base and the orientation of technological and institutional change in such a manner as to ensure the attainment and continued satisfaction of human needs for the present and future generations. Such sustainable development (in the agriculture, forestry and fisheries sector) conserves land, water, plant and animal genetic resources, is environmentally non-degrading, technically appropriate, economically viable and socially acceptable. So, to achieve sustainable agriculture development the optimum use of natural resources, human resources, capital resources and technical resources are required.

In India the crop yield is heavily dependent on rain which is the main reason for the declining growth rate of agriculture sector. These uncertainties hit the small farmers and laborers worst which are usually leading a hand to mouth life. Therefore something must be done to support farmers and sufficient amount of water and electricity must be supplied to them as they feel insecure and continue to die of drought, flood, and fire. India is the second largest country of the world in terms of population; it should realize it is a great resource for the country. India has a huge number of idle people. There is a need to find ways to explore their talent and make the numbers contribute towards the growth. Especially in agriculture passive unemployment can be noticed.

The sustainable development in India can also be achieved by full utilization of human resources.A large part of poor population of the country is engaged in agriculture, unless we increase their living standard, overall growth of this country is not possible. If we keep ignoring the poor, this disparity will keep on increasing between classes. Debt traps in country are forcing farmers to commit suicides. People are migrating towards city with the hope of better livelihood but it is also increasing the slum population in cities. Therefore rural population must be given employment in their areas and a chance to prosper. India has been carrying the tag of "developing" country for quite long now; for making the move towards "developed" countries we must shed this huge dependence on agriculture sector.

Conclusion

The agricultural technology needs to move from production oriented to profit oriented sustainable farming. The conditions for development of sustainable agriculture are becoming more and more favorable. New opportunities are opening the eyes of farmers, development workers, researchers and policy makers like agri related businesses, dairy farming, poultry farming, castle farming and fisheries. Now the time is to see the potential and importance of these practices not only for their economic interest but also as the basis for further intensification and ecological sustainability. To conclude, a small-farm management to improve productivity, profitability and sustainability of the farming system will go a long way to ensure all round sustainability.

References

Bhattacharya, B.B. (2003). "Trade Liberalization and Agricultural Price Policy in India since Reforms", *Indian Journal of Agricultural Economics*, Vol.58, No.3

Braun, Joachim von, Gulati, A., Hazell, P., Mark W. Rosegrant and Ruel, Marie (2005). Indian Agriculture and Rural Development- Strategic Issues and Reform Options.

Dev, S. Mahendra (2008). Inclusive Growth in India, Agriculture, Poverty and Human Development, Oxford University Press, New Delhi.

Evenson, R.E.; Pray, C. and Rosegrant, M.W. (1999). Agricultural Research and Productivity Growth in India. Research Report No 109. International Food Policy Research Institute, Washington, D.C.

GOI (2007). Report of the Steering Committee on Agriculture for 11th Five Year Plan, Yojana Bhavan, New Delhi.

GOI (2007a). Agricultural Strategy for the Eleventh Plan: Concerns and Way ahead, Yojana Bhavan, New Delhi.

Gulati, Ashok (2009), "Emerging Trends in Indian Agriculture: What can we learn from these?" Prof. Dayanath Jha Memorial Lecture, National Centre for Agricultural Economics and Policy Research, New Delhi.

Gulati, Ashok. Meinzen-Dick., Ruth, and Raju, K.V. (2005) Institutional Reforms in India Irrigation, Sage Publication.

International Journal of Management Research and Technology "Productivity and Sustainability in Agriculture: An Application of LPP Model", Vol.2 No.2 July-Dec 2008.

Kumar, Praduman and Mittal.,Surabhi (2006). "Agricultural Productivity Trends in India: Sustainability Issues" Agricultural Economic Research Review. Volume 19, pp 71-88.

Kushwaha Niru (2003). Environment, Sustainable Development and Rural Poverty in India. Ph.D. Thesis, M.J.P. Rohilkhand University, Bareilly Ch. 4.

Mishra, V.N. and Rao, Govinda (2003). Trade Policy, Agricultural Growth and Rural Poor: Indian Experience, 1978-79 to 1999-00, Economic and Political Weekly, October 25, 2003 [13]. Mittal, Surabhi (2006a). "Past Trend and Projections of Demand and Supply for Major Food Crops in India". Background paper prepared for Planning Commission, Government of India. June, 2006.

Promoting Sustainable Agriculture in India, available at: http://www.articlesbase.com/agriculture-articles/promoting-sustainable-agriculture-in-india-2132445.html#ixzz1FWsPazLc

Rao, Hanumantha., C.H. (2003). "Reform Agenda for Agriculture", Economic and Political Weekly, Feb 15, 2003.

Singh, Panjab (2002). Agricultural Policy-Vision 2020. Planning Commission. http://planningcommission.nic.in/reports/genrep/bkpap2020/24_bg2020.pdf

Virmani, Arvind, (2004). Accelerating growth and Poverty Reduction: A Policy Framework for India's Development. Academic Foundation, New Delhi.

Vyas V. S. (2003). India's Agrarian Structure, Economic Policies and Sustainable Development, Academic Foundation Publishers, New Delhi.

17

Environmental Policies, Laws and Legislation

Usha Mishra[1] *and Priyanka Mishra*[2]

[1]H.O.D., M.O.M. Govt.Women Polytechnic, Jabalpur
[2]Research Scholar, Rani Durgavati Vishva Vidyalaya, Jabalpur

ABSTRACT

A diverse developing society such as ours provides numerous challenges in the economic, social, political, cultural, and environmental arenas. All of these coalesce in the dominant imperative of alleviation of mass poverty, reckoned in the multiple dimensions of livelihood security, health care, education, empowerment of the disadvantaged, and elimination of gender disparities.

The present national policies for environmental management are contained in the, the; and the. Some sector policies such as the and have also contributed towards environmental management. All of these policies have recognized the need for sustainable development in their specific contexts and formulated necessary strategies to give effect to such recognition. The National Environment Policy seeks to extend the coverage, and fill in gaps that still exist, in light of present knowledge and accumulated experience. It does not displace, but builds on the earlier policies. National Forest Policy, 1988 National Conservation Strategy and Policy Statement on Environment and Development, 1992 Policy Statement on Abatement of Pollution, 1992 National.

The National Environment Policy is intended to be a guide to action: in regulatory reform, programmes and projects for environmental conservation; and review and enactment of legislation, by agencies of the Central, State, and Local Governments. The policy also seeks to stimulate partnerships of different stakeholders, i.e. public agencies, local communities, academic and scientific institutions, the investment community, and international development partners, in harnessing their respective resources and strengths for environmental management. The dominant theme of this policy is that while conservation of environmental resources is necessary to secure livelihoods and well-being of all, the most secure basis for conservation is to ensure that people dependent on particular resources obtain better livelihoods from the fact of conservation, than from degradation of the resource.

Through this paper we are likely to submit our thoughts in present situation of the society.

Keywords: *Environment, Polices, Laws.*

Introduction

Word "environment" is most commonly used describing "natural" environment and means the sum of all living and non-living things that surround an organism, or group of organisms. Environment includes all elements, factors, and conditions that have some impact on growth and development of certain organism. Environment includes both biotic and a-biotic factors that have influence on observed organism. A-biotic factors such as light, temperature, water, atmospheric gases combine with biotic factors (all surrounding living species). Environment often changes after some time and therefore many organisms have ability to adapt to these changes. However tolerance range is not the same with all species and exposure to environmental conditions at the limit of an certain organism's tolerance range represents environmental stress.

Environment is the immediate surrounding space around man. It includes a biotic and biotic component. So environment not only means our environment but also varies of issues connected with human activity and its impact on natural resources. It has been observed that in recent past far reaching changes have taken place. Man has played a key role in modifying the environment in his constant efforts towards improving the standard of living.

Environment is defined as the surroundings in which the organism lives. The environment may be the physical environment, the chemical environment or the biological environment. Organisms are dependent on the environment to fulfill their needs; man is also constantly interacting with the environment in order to fulfill his needs. These needs include the basic needs of oxygen, food and shelter in addition to the social needs like entertainment, medicines, etc. The things that man requires for his survival and comfort are called the resources. The environment is a reservoir of resources. Maintaining the natural resources of the environment and their careful use is called conservation. The conservation of environment involves the conservation of the natural resources.

A healthy environment is an absolute necessity for the well-being of all organisms, including man. All our needs, big and small are being met by the environment. However, man having reached the pinnacle of evolution is trying to bring about changes in the environment to suit his convenience. Unfortunately, this convenience is temporary. In the long run, man is losing out on a healthy environment.

Causes, Sources and Effects of Environmental Pollution

Environmental pollution is "the contamination of the physical and biological components of the earth/atmosphere system to such an extent that normal environmental processes are adversely affected""Pollution is the introduction of contaminants into the environment that cause harm or discomfort to humans or other living organisms, or that damage the environment" which can come "in the form of chemical substances, or energy such as noise, heat or light". "Pollutants can be naturally occurring substances or energies, but are considered contaminants when in excess of natural levels."

Pollution is "the addition of any substance or form of energy (e.g., heat, sound) to the environment at a rate faster than the environment can accommodate it by dispersion, breakdown, recycling, or storage in some harmless form.

Environmental Pollutants

Environmental pollutants are constituent parts of the pollution process. They are the actual "executing agents" of environmental pollution. They come in gaseous, solid or liquid form. It is interesting to note that, as of 1990, there were around 65,000 different chemicals in the marketplace, *i.e.* potential environmental pollutants that were to be released into air, water and land on a regular basis.

Renowned author Miguel A. Santos identifies at least three general characteristics of environmental pollutants9

I. Pollutants don't recognize boundaries, *i.e.* they are trans boundary;
II. Many of them can't be degraded by living organisms and therefore stay in the ecosphere for many years; and
III. They destroy biota and habitat.

We know that decomposition of pollutants can occur either biologically or physic chemically.

Types of Pollutant

Pollutants are divided into two catrigories- Biodegradable and Non-biodegradable.

Biodegradable Pollution

Pollution that is rendered harmless by natural processes and so causes no permanent harm biodegradable pollution is mainly organic wastes and stuff like banana peels, hamburger buns, etc. These things are decomposed by bacteria, fungi and other microbes like that.

Non-biodegradable Pollution

Pollution that accumulates in the environment and may appear in the food Chain.

Types of Pollution

Generally speaking, there are many types of pollution but the most important ones are:

I. Air pollution
II. Water pollution
III. Soil pollution (contamination)

Some of the most notable air pollutants are sulfur dioxide, nitrogen dioxide, carbon monoxide, ozone, volatile organic compounds (VOCs) and airborne particles, with radioactive pollutants probably among the most destructive ones (specifically when produced by nuclear explosions).

Water pollutants include insecticides and herbicides, food processing waste, pollutants from livestock operations, volatile organic compounds (VOCs), heavy metals, chemical waste and others. Some soil pollutants are: hydrocarbons, solvents and heavy metals.

Sources of Environmental Pollution

Fossil Fuel Sources of Environmental Pollution

Industry -I. Power-generating plants, II. Petroleum refineries, III. Petrochemical plants, IV. Production and distribution of fossil fuels, V. Other manufacturing facilities

Transport

I. Road transport (motor vehicles),II. Shipping industry, III. Aircraft

Environmental Pollution Effects: Bhopal Gas Case study

Bhopal Pollution Disaster: What Exactly Happened in Bhopal in 1984?

On the night of December 2 1984, a deadly dosage of the poisonous methyl isocyanate gas leaked from a tank of the pesticides plant in Bhopal owned by the US Company Union Carbide. The gases swirled across the ground into nearby bustles (slums), killing Bhopal's poorest inhabitants in their sleep, burning the eyes and lungs of survivors, and causing nearly two decades of deaths, injuries and ill health. Estimates of the number of people killed on the first night range from an official figure of 3,000 to as many as 7,000 or 8,000 based on the number of kafans, or shrouds, ordered by religious organizations for wrapping the dead). Since then, at least another 10,000 to 15,000 of those affected have died.

Environmental Protection: The Laws

The constitutional provisions are backed by a number of laws – acts, rules, and notifications. The EPA (Environment Protection Act), 1986 came into force soon after the Bhopal Gas Tragedy and is considered an umbrella legislation as it fills many gaps in the existing laws. Thereafter a large number of laws came into existence as the problems began arising, for example, Handling and Management of Hazardous Waste Rules in 1989.

Following is a list of the environmental legislations that have come into effect General, Forest and wildlife, Water, Air.

General

- ☆ The Environment (Protection) Act16 authorizes the central government to protect and improve environmental quality, control and reduce pollution from all sources, and prohibit or restrict the setting and/or operation of any industrial facility on environmental grounds.
- ☆ The Environment (Protection) Rules 17 lay down procedures for setting standards of emission or discharge of environmental pollutants.

- ☆ The objective of Hazardous Waste (Management and Handling) Rules 18is to control the generation, collection, treatment, import, storage, and handling of hazardous waste.
- ☆ The Manufacture, Storage, and Import of Hazardous Rules define the terms used in this context, and sets up an authority to inspect, once a year, the industrial activity connected with hazardous chemicals and isolated storage facilities.
- ☆ The Manufacture, Use, Import, Export, and Storage of hazardous Micro-organisms/Genetically Engineered Organisms or Cells Rules were introduced with a view to protect the environment, nature, and health, in connection with the application of gene technology and microorganisms.
- ☆ The Public Liability Insurance Act and Rules and Amendment, was drawn up to provide for public liability insurance for the purpose of providing immediate relief to the persons affected by accident while handling any hazardous substance.
- ☆ The National Environmental Tribunal Act has been created to award compensation for damages to persons, property, and the environment arising from any activity involving hazardous substances.
- ☆ The National Environment Appellate Authority Act has been created to hear appeals with respect to restrictions of areas in which classes of industries etc. are carried out or prescribed subject to certain safeguards under the EPA.
- ☆ The Biomedical waste (Management and Handling) Rules is a legal binding on the health care institutions to streamline the process of proper handling of hospital waste such as segregation, disposal, collection, and treatment.
- ☆ The Environment (Siting for Industrial Projects) Rules, 1999 lay down detailed provisions relating to areas to be avoided for siting of industries, precautionary measures to be taken for site selecting as also the aspects of environmental protection which should have been incorporated during the implementation of the industrial development projects.
- ☆ The Municipal Solid Wastes (Management and Handling) Rules, 2000 apply to every municipal authority responsible for the collection, segregation, storage, transportation, processing, and disposal of municipal solid wastes.
- ☆ The Ozone Depleting Substances (Regulation and Control) Rules have been laid down for the regulation of production and consumption of ozone depleting substances. The Batteries (Management and Handling) Rules, 2001 rules shall apply to every manufacturer, importer, re-conditioner, assembler, dealer, auctioneer, consumer, and bulk consumer involved in the manufacture, processing, sale, purchase, and use of batteries or components so as to regulate and ensure the environmentally safe disposal of used batteries.

- ☆ The Noise Pollution (Regulation and Control) (Amendment)Rules lay down such terms and conditions as are necessary to reduce noise pollution, permit use of loud speakers or public address systems during night hours (between 10:00 p.m. to 12:00 midnight) on or during any cultural or religious festive occasion.
- ☆ The Biological Diversity Act is an act to provide for the conservation of biological diversity, sustainable use of its components, and fair and equitable sharing of the benefits arising out of the use of biological resources and knowledge associated with it.

Forest and Wildlife

- ☆ The Indian Forest Act and Amendment, 1984, is one of the many surviving colonial statutes. It was enacted to 'consolidate the law related to forest, the transit of forest produce, and the duty leviable on timber and other forest produce'.
- ☆ The Wildlife Protection Act, Rules 1973 and Amendment 1991 provides for the protection of birds and animals and for all matters that are connected to it whether it be their habitat or the waterhole or the forests that sustain them.
- ☆ The Forest (Conservation) Act and Rules, provides for the protection of and the conservation of the forests.

Water

- ☆ 1882 - The Easement Act allows private rights to use a resource that is, groundwater, by viewing it as an attachment to the land. It also states that all surface water belongs to the state and is a state property.
- ☆ 1897 - The Indian Fisheries Act establishes two sets of penal offences whereby the government can sue any person who uses dynamite or other explosive substance in any way (whether coastal or inland) with intent to catch or destroy any fish or poisonous fish in order to kill.
- ☆ 1956 - The River Boards Act enables the states to enroll the central government in setting up an Advisory River Board to resolve issues in inter-state cooperation.
- ☆ 1970 - The Merchant Shipping Act aims to deal with waste arising from ships along the coastal areas within a specified radius.
- ☆ 1974 - The Water (Prevention and Control of Pollution) Act establishes an institutional structure for preventing and abating water pollution. It establishes standards for water quality and effluent. Polluting industries must seek permission to discharge waste into effluent bodies. T he CPCB (Central Pollution Control Board) was constituted under this act.
- ☆ 1977 - The Water (Prevention and Control of Pollution) Cess Act provides for the levy and collection of cess or fees on water consuming industries and local authorities.

- ☆ 1978 - The Water (Prevention and Control of Pollution) Cess Rules contains the standard definitions and indicate the kind of and location of meters that every consumer of water is required to affix.
- ☆ 1991 - The Coastal Regulation Zone Notification puts regulations on various activities, including construction, are regulated. It gives some protection to the backwaters and estuaries.

Air

- ☆ 1948 – The Factories Act and Amendment in 1987 was the first to express concern for the working environment of the workers. The amendment of 1987 has sharpened its environmental focus and expanded its application to hazardous processes.
- ☆ 1981 - The Air (Prevention and Control of Pollution) Act provides for the control and abatement of air pollution. It entrusts the power of enforcing this act to the CPCB.
- ☆ 1982 - The Air (Prevention and Control of Pollution) Rules defines the procedures of the meetings of the Boards and the powers entrusted to them.
- ☆ 1982 - The Atomic Energy Act deals with the radioactive waste.
- ☆ 1987 - The Air (Prevention and Control of Pollution) Amendment Act empowers the central and state pollution control boards to meet with grave emergencies of air pollution.
- ☆ 1988 - The Motor Vehicles Act states that all hazardous waste is to be properly packaged, labeled, and transported.

Role of Indian Judiciary in Environmental Protection

Judicial and Quasi Judicial Bodies for Environmental Protection are the Courts or Authorities established under various environmental Protection Legislations for interpretation and effective implementation of these statutes.

Supreme Court of India

I. Public Interest Litigation (PIL)

II. Special Leave Petition

Public Interest Litigation (PIL)

A Public Interest Litigation (PIL) can be filed in any High Court or directly in the Supreme Court. It is not necessary that the petitioner has suffered some injury of his own or has had personal grievance to litigate. PIL is a right given to the socially conscious member or a public spirited NGO to espouse a public cause by seeking judicial for redressal of public injury. Such injury may arise from breach of public duty or due to a violation of some provision of the Constitution. Public interest litigation is the device by which public participation in judicial review of administrative action is assured. It has the effect of making judicial process little more democratic.

M.C. Mehta vs. State of Orissa

A writ petition was filed to protect the health of thousands of innocent people living in Cuttack and adjacent areas who were suffering from pollution from sewage being caused by the Municipal Committee Cuttack and the SCB Medical College Hospital, Cuttack.

The main contention of the petitioner was that the dumping of untreated waste water of the hospital and some other parts of the city in the Taladanda canal was creating health problems in the city. The State, on the other hand contended that a central sewerage system had been installed in the hospital and that there is no sewage flow into the taladanda canal as alleged. Further, it was asserted that the State had not received any information relating to either pollution or of epidemic of water borne diseases caused by contamination of the canal. Also, the health department shrugged off the responsibility for supply of drinking water and passed the buck to the Municipality which refuted the contentions of carelessness and callousness.

The Court reprimanded the authorities and directed the government to immediately act on the matter. Also, the court recommended setting up of a committee to take steps to prevent and control water pollution and to maintain wholesomeness of water meant for human consumption amongst other things. A responsible Municipal Council is constituted for the precise purpose of preserving public health. Provision of proper drainage system in working conditions cannot be avoided by pleading financial inability.

M.C Mehta vs. Union of India

AIR 1997 SC 734

On the eve of his retirement, J. Kuldip Singh delivered the judgment in the Taj Trapezium case, culminating a long and arduous battle fought by M. C Mehta for over a decade. The case was first placed in 1984, wherein the petitioner warned of damage to the Taj Mahal from air pollutants from the Mathura refinery.

It was alleged by the petitioner that the suplphur dioxide emitted by the Mathura refinery and the industries when combined with Oxygen-with the aid of moisture-in the atmosphere forms suplphuric acid which has a corroding effect on the gleaming white marble of the Taj. Industrial/Refinery emissions, brick-kilns, vehicular traffic and generator sets are primarily responsible for polluting the ambient air around Taj Trapezium. The petition states that the white marble has yellowed and blackened in places. It is inside the Taj that the decay is more apparent. Yellow pallor pervades the entire monument. In places the yellow hue is magnified by ugly brown and black spots. Fungal deterioration is worst n the inner chamber where the original graves of Shah-Jahan and Mumtaz mahal lie.The Court observed that the Taj, apart from being cultural heritage, is an industry by itself, More than two million tourists visit the Taj every year. It is a source of revenue for the country.

Various orders were passed by the Court. The Court created a Taj Trapezium which consisted of 10, 400 sq. Kms in the shape of a trapezium to regulates activities in relation to air pollution.Industries were asked to shift to eco friendly fuel and use lessen the use of diesels generators, and asked the State to improve power supply the city. Tanneries operating from Agra were asked to shift from the Trapezium.

Conclusion and Suggestions

Environment is an important part of human life and a healthy environment is a must for human existence. Therefore, it is crucial that we take care of our surroundings and help nature maintain ecological balance so that we could hand over to the coming generations the environment as we found it, if not any better. In the recent past there has been a lot of damage to the ecology. Air, water and soil have been polluted and there appears to be no decisive end to it. The scientific advancement and rapid industrialization has taken its toll.

Nowadays protection of the environment is very important as the world is moving into a new era without considering any of the major problems of pollution with rapid industrialization. The best way to protect the environment is conservation. Conservation is the philosophy and policy of managing the environment to assure adequate supplies of natural resources for future as well as present. Tropical forests are being destroyed at an ever-increasing rate. Estimates of the extent and rate of loss vary, but it appears that nearly half of the world tropical forests already have been lost, and the remainder will all but disappear in the next two to three decades. The loss is incalculable. These forests provide habitat for an estimated half of the world plant and animal species, provide water and fuel for much of the world population, and influence regional and global climate. Commercial logging, clearance for agriculture.

The Courts in India have played a distinguishing role in gradually enlarging the scope of a qualitative living by applying various issues of environmental protection. Consequently, activities posing a major threat to the environment were curtailed so as to protect the individual's inherent right to wholesome environment. Art 21 has been relied in the plethora of cases, although certain cases have incorporated a wider perspective of the Constitution. Hence, the Supreme Court of India, apart from being environment friendly, has given birth to a wide range of doctrines and principles have in turn been adopted and implemented throughout the country.

The suggestions made by the Supreme Court in *A.P. Pollution Control Board v. M.V.Nayudu*, for the improvement of the adjudicatory machinery under the various environmental laws should be implemented by the Government. The main burden of these suggestions is that in all environmental courts, tribunals and appellate authorities, there should be a judge of the rank of a High Court or a Supreme Court, sitting or retired, and a scientist or a group of scientists of high ranking and experience so as to help a proper and fair adjudication of disputes relating to environment and protection. If implemented, this would go a long way in securing justice to the needy.

Any law is as good as the implementation. The implementation mechanism must be foolproof and effective. There must be an effective monitoring mechanism. The bottom line is that we must have individuals of integrity with a strong value base and deep commitment if laws are to be effectively implemented.

"What's the use of a fine house if you haven't got a tolerable planet to put it on?"

— Henry David Thoreau, Familiar Letters

18

पर्यावरण नीतियां, कानून एवं विधायन भारत के विशेष संदर्भ में (विश्लेषणात्मक अध्ययन)

मयूर शिखा अग्रवाल

प्राध्यापक विधि, प्रभारी प्राचार्य, पूर्व डीन लॉ फैकल्टी, बरतकतउल्ला वि.वि. भोपाल
शास. राज्य स्तरीय विधि स्नातकोत्तर महाविद्यालय, भोपाल (म.प्र.)

सार-संक्षेप

उच्च कोटि की प्रौद्योगिकी से औद्योगिकी करण की प्रक्रिया का शुभारंभ हुआ और औद्योगिक मानव का आर्थिक मानव के रूप में कायाकल्प हो गया। उद्योगों से अच्छी कमाई की लौलुपता में उसने प्राकृतिक संसाधनों का अविवेकपूर्ण दोहन करना आरंभ कर दिया। प्राकृतिक संसाधनों के अतिदोहन से अनेक पर्यावरणीय और परिस्थितिकीय समस्याओं का उद्भव हुआ।

मनुष्य के आचरण को नियंत्रित करने के लिए अनेक पर्यावरण संरक्षण संबंधी कानून बनाने की आवश्यकता आई।

संकेत शब्दः पर्यावरण, कानून, भारत

पर्यावरण संरक्षण को लेकर भारत सदा सजग रहा है। हमारे देश में पर्यावरण के अनुकूल एक समृद्ध संस्कृति रही है। मनुष्य व पर्यावरण के संबंध को हम तीन निम्नलिखित अवस्थाओं में बांट सकते हैं:

1. **आदिमानव के साथ संबंधः** मनुष्य पर्यावरण में पैदा हुआ, जब मानव ने प्रथम शिशु के रूप में आंखे खोली होगी तब उसके चारों ओर वनस्पतियां, जलाशय, पवन, नीला गगन, चहचहाती चिड़ियां, चमकती धूप, प्रथम साथी के रूप में मिले होंगें। प्रकृति का सहारा लेकर उसकी गोद में पलकर उसने सभ्य मानव तक की यात्रा की।
2. **सभ्य मानव के रूप में संबंधः** सभ्य मानव वह था जो प्रकृति से अपनी आवश्यकता के लिए जितना लेता उससे कई गुना लौटाता, यही वह काल था जब मनुष्य व पर्यावरण एक दूसरे के पूरक थे ना कि विनाशक। यह मानव व पर्यावरण में सहभागिता का परस्पर परावलम्बन का काल था।
3. **प्रौद्योगिक मानव के रूप में संबंधः** उन्नसवी शताब्दी के पूर्वाद्ध तक का प्रौद्योगिक मानव पर्यावरण का विनाशक नहीं था बल्कि उसने पर्यावरण व परिस्थिति तंत्र को और भी व्यवस्थित किया था। यातायात के लिए सड़के, सिंचाई के साधन, खेती को उन्नत करने हेतु नहरें, बांस से कागज का निर्माण, पढ़ाई–लिखाई के विस्तार हेतु आदि तैयार करके।

मनन–चिंतन के सुअवसर के दो परिणाम सामने आयेः

1. सांस्कृतिक पर्यावरण का निर्माण
2. उच्च कोटि की प्रौद्योगिकी,

उच्च कोटि की प्रौद्योगिकी से औद्योगिकी करण की प्रक्रिया का शुभारंभ हुआ और औद्योगिक मानव का आर्थिक मानव के रूप में कायाकल्प हो गया। उद्योगों से अच्छी कमाई की लौलुपता में उसने प्राकृतिक संसाधनों का अविवेकपूर्ण दोहन करना आरंभ कर दिया। प्राकृतिक संसाधनों के अतिदोहन से अनेक पर्यावरणीय और परिस्थितिकीय समस्याओं का उद्भव हुआ।

मनुष्य के आचरण को नियंत्रित करने के लिए अनेक पर्यावरण संरक्षण संबंधी कानून बनाने की आवश्यकता आई।

स्वतंत्रता पूर्व पर्यावरण कानून

स्वतंत्रता के पूर्व हमारा संविधान अस्तित्व में नहीं आया था। परन्तु पर्यावरण की तरफ विधिक स्तर पर ध्यान दिया जाता था।

1. भारतीय दण्ड संहिता 1860 संबंधित धाराऐं – 268, 290, 291, 426, 430, 431, 432, 278.
2. जल प्रदूषण से संबंधित कानून – दि नार्थ कैनाल एण्ड ड्रेनेज एक्ट 1873, दि ऑब्स्ट्रक्शन ऑफ फेयरवेज एक्ट, 1881, इंडियन फिशरीज एक्ट 1897.

3. वायु प्रदूषण से संबंधित कानून – दि ओरिएन्टल गैस कंपनी एक्ट 1857, दि एक्सपलोसिव एक्ट 1908, इंडियन वायलर्स एक्ट1923, दि मोटर व्हीकल एक्ट, 1938.
4. वन्य जीव संरक्षण संबंधित कानून
5. कीटाणु नाशक कृषि संरक्षण संबंधित कानूनदि पायजन एक्ट, 1911, दि मैसूर डिस्ट्रेक्टिव एण्ड पेस्ट एक्ट, 1917, दि आंधप्रदेश एग्रीकल्चर पेस्ट एण्ड डिजीज एक्ट, 1919, दि केरल एग्रीकल्चर पेस्ट एण्ड डिजीज एक्ट, 1917.

स्वतंत्रता पश्चात के प्रमुख कानून

भारतीय संविधान एवं पर्यावरण

पर्यावरण को संरक्षित रखने की दृष्टि से संविधान में निम्न प्रावधान है।

अनुच्छेद 21 – स्वच्छ पर्यावरण में जीने का मौलिक अधिकार ।

अनुच्छेद 47 – लोक स्वास्थ्य में सुधार हेतु राज्य के नीति निर्देशक तत्व।

अनुच्छेद 48ए –वन व वन्य जीवों की रक्षा हेतु नीति निर्देशक तत्व

अनुच्छेद 51A(g) – प्राकृतिक पर्यावरण की रक्षा का नागरिकों का कर्त्तव्य।

अनुच्छेद 32 एवं 226 – संवैधानिक उपचार।

अनुच्छेद 253 – संसद को विधि बनाने की शक्ति।

पर्यावरण और विनिर्दिष्ट विधायन

भारत में पर्यावरण संरक्षण तथा प्रदूषण निवारण और नियंत्रण के लिए जो विनिर्दिष्ट विधायन (Specific Legislation) लागू हैं, उन्हें दो भागों में बांटकर देखा जा सकता है। (1) केन्द्रीय विधायन (2) राज्य विधायन।

केन्द्रीय विधायन (Central Legislation)

1. पर्यावरण (संरक्षण) अधिनियम 1986
2. जल (प्रदूषण निवारण एवं नियंत्रण) अधिनियम 1974
3. वायु (प्रदूषण निवारण एवं नियंत्रण) अधिनियम 1981
4. भारतीय वायलर अधिनियम, 1923
5. भारतीय वन अधिनियम, 1927
6. वन (संरक्षण) अधिनियम, 1972

7. वन्य जीव (संरक्षण) अधिनियम, 1972
8. वन्य जीव (स्टाक घोषणा) केन्द्रीय नियम, 1973
9. कीटनाशी अधिनियमय, 1968
10. विष अधिनियम, 1919
11. लोक दायित्व बीमा अधिनियम 1991
12. परिसंकटमय अपशिष्ट (प्रबंधन और व्यवहार) नियम, 1991
13. जैव चिकित्सा अपशिष्ट नियमावली 1998
14. जैव विविधता अधिनियम, 2002
15. राष्ट्रीय हरित अधिकरण अधिनियम, 2010

उपरोक्त प्रमुख अधिनियमों के संबंधित नियम भी बनाये गए।

राज्यों ने अपने राज्य में पर्यावरण संरक्षण तथा प्रदूषण निवारण और नियंत्रण के लिए पृथक से अधिनियम व नियम पारित कर लागू किये हैं।

पर्यावरण संरक्षण अधिनियम, 1986

पर्यावरण संरक्षण संबंधित स्टॉकहोम सम्मेलन, 1972 संयुक्त राष्ट्र के तत्वाधान में आयोजित हुआ जिसमें विश्व के सभी भागों ने भाग लेकर पर्यावरण प्रदूषण के बढ़ते हुए खतरे को देखते हुए अपनी चिंता व्यक्त की। इस सम्मेलन में हमारी तत्कालीन प्रधानमंत्री ने भाग लिया। इस सम्मेलन में सभी राष्ट्र ने सर्वसम्मति से पर्यावरण सुरक्षा से संबंधित विधियां पारित की हैं। तदानुसार भारतीय संसद द्वारा पर्यावरण (संरक्षण) अधिनियम, 1986 पारित किया गया।

उद्देश्य

अधिनियम के मुख्य उद्देश्य निम्नलिखित हैं:

1. मानवीय पर्यावरण पर स्टॉकहोम सम्मेलन में लिए गए निर्णयों को साकार रूप देना।
2. मानव प्राणियों जीवों, पादपों एवं संपत्तियों का परिसंकट से बचाना।
3. पर्यावरण संरक्षण हेतु सामान्य एवं व्यापक विधि अधिनियमित करना।
4. मानवीय पर्यावरण सुरक्षा एवं स्वास्थ्य को खतरा उत्पन्न करने वालों के लिए निरोधात्मक दण्ड की व्यवस्था करना।

5. विद्यमान विधियों के अधीन विभिन्न अभिकरणों के कार्यकलाप के बीच समन्वय स्थापित करना।

इस अधिनियम के अंतर्गत सिर्फ यह कहा जा सकता है कि पर्यावरण सुरक्षा का दायित्व केवल सरकार का ही नहीं अपितु यह प्रत्येक नागरिक का कर्त्तव्य है कि वह पर्यावरण प्रदूषण के घातक परिणामों के प्रति सचेत रहे तथा इसके विरूद्ध लड़ने हेतु संकल्प लेकर सरकार द्वारा लिए गए उपायों में मदद करें तथा पर्यावरण में अपना सक्रिय योगदान दें।

एम. निजामुद्दीन बनाम केम्प्लास्ट सनमार लिमिटेड तथा अन्य (A.I.R. 2010 सु.को. 1765) के प्रकरण में पर्यावरण संरक्षण अधिनियम की धारा 25 तथा इस अधिनियम के अधीन नियमों का नियम 5 और समुद्र तटीय विनियमन जोन अधिसूचना, 1991 के पैरा 3 ;3) की व्याख्या करते हुए उच्चतम ने विनिश्चित किया कि यदि प्रत्यर्थी को तटीय विनियमन जोन के अधीन संयंत्र लगाने की अनुमति पास हो चुकी है जिसे राज्य सरकार केन्द्रीय पर्यावरण एवं वन मंत्रालय की मंजूरी भी प्राप्त हो गई हो तो इन दोनों में सन् 1998 तथा 2002 के पश्चातवर्ती परिवर्तन का विपरीत प्रभाव नहीं पड़ेगा और पर्यावरण की वैद्यता अवैध नही होगी।

पर्यावरण संरक्षण अधिनियम, 1986 की धारा 6 और 25 के अधीन केन्द्रीय सरकार ने नियमावली बनाई जो दिनांक 19 नवम्बर 1986 से प्रभावित हुई। यह नियमावली अधिनियम के उपबंधों के क्रियान्वित करने और पर्यावरण से संबंधित विभिन्न समस्याओं का समाधान करने की दिशा में बहुत महत्वपूर्ण सिद्ध; हुई।

पर्यावरण कानूनों के प्रवर्तन में न्यायपालिका की भूमिका

लोक कल्याणकारी राज्य में जनसमस्याओं को दूर करने के लिये न्यायपालिका ने न्यायिक सक्रियता दिखाते हुए पर्यावरण संरक्षण के लिए महत्वपूर्ण निर्णय दिये, रूरल लिटिगेशन एण्ड एनवायरमेंट केन्द्र एवं देवकीनन्दन पाण्डे बनाम भारत संघ ए.आई.आर. 1985 एस.सी. 652 एवं उत्तरप्रदेश राज्य AIR 1987 एस.सी. 395, ये दोनों वाद देहरादून की चूना पत्थर खदानों की खुदाई संबंधी वाद है। इनसे आसपास के रहने वाले निवासियों को पर्यावरण प्रदूषण से हानि पहुंच रही थी।

लोकहितवाद पर न्यायालय ने जांच समिति की नियुक्ति की व समिति की रिपोर्ट के आधार पर इन पत्थर खानों की खुदाई का काम रोकने के आदेश दिये।

समाजसेवी एस.सी. मेहता द्वारा दायर विभिन्न पर्यावरण संरक्षण हेतु वादों में उच्चतम न्यायालय वे महत्वपूर्ण निर्णय दिये क्रमशः इस प्रकार हैं:

एम.सी. मेहता बनाम भारत संघ ए.आई.आर. (1980) एस.सी. 1115

(गंगा प्रदूषण वाद-I)

यह जनहित वाद विधि द्वारा आरोपित लोक कर्तव्य का निर्वहन नहीं कर रहे , नगर निगम और प्रदूषण नियंत्रण बोर्ड के विरूद्ध दायर किया गया था। न्यायालय ने निर्देश जारी किये उन सभी नगरपालिकाओं एवं नगर निगमों के लिए जहां से होकर गंगा नदी बहती है।

1. कानपुर नगरपालिका को यह निर्देश दिया कि 6 माह में जल परिषद को जल प्रदूषण रोकने का प्रस्ताव भेजे।
2. दूध की डेयरी को शहर के बाहर ले जाया जाये।
3. श्रमिक कालोनी में सीवर लाईन बनायी जाए।
4. निर्धन लोगों के लिए सार्वजनिक शौचालय एवं मूत्रालय बनाये जायें।
5. मरे हुए पशुओं और मानव लाशों को गंगा में फेकने से रोका जाए।
6. नये कारखानों को स्थापित करने के लाइसेन्स तब तक न दिये जाये जब तक वे प्रदूषण रोकने के लिए पर्याप्त उपमान न लगा लें।

एम.सी. मेहता बनाम भारत संघ ए.आई.आर. (1988) एस.सी. 1037

(गंगा प्रदूषण वाद-II)

कानपुर के चमड़े कारखाने से निकलने वाले मलवे से गंगा का पानी प्रदूषित हो रहा था। कारखाना स्वामियों द्वारा अपशिष्ट जल के शुद्धिकरण के संयंत्र स्वामियों द्वारा नहीं लगाए गए थे, उनका तर्क था संयंत्र खर्चीले हैं एवं कारखाना बंद करना बेरोजगारी व राजस्व को हानि पहुचाने वाला होगा। न्यायपालिका ने कारखाना स्वामियों के तर्क अस्वीकार करते हुए निर्णय दिया कि लाखों लोगों के जीवन स्वास्थ्य और वायुमण्डल की देखभाल करें।

वेल्लोर सिटीजन्स वेलफेयर फोरम बनाम भारत संघ 1965 एस.सी.सी. 647

तमिलनाडु के चमड़ा व अन्य व्यवसायों से निकलने वाले अशुद्ध मलबे से लगभग 13 गॉवों के 350 कुओं का पानी प्रदूषित हो गया था। न्यायालय ने प्रत्येक चमड़ा कारखाने पर 10,000/— रू. का प्रदूषण दण्ड लगा कर प्रदूषण संरक्षण फण्ड का निर्माण किया। इस फण्ड से पीड़ितों को प्रतिकर दिलाने के निर्देश दिये।

राष्ट्रीय हरित न्यायाधिकरण (National Green Tribunal)

पर्यावरण संबंधी विवादों के शीघ्र निपटारे तथा पर्यावरण से संबंधित कानूनों के प्रभावी क्रियान्वपयन को ध्यान में रखकर सरकार द्वारा एक राष्ट्रीय हरित न्यायाधिकरण (National

Green Tribunal NGT) की स्थापना की गई है, जिसका मुख्यालय भोपाल में बनाया गया है। गौरतलब है कि इस तरह के न्यायाधिकरण का गठन करने वाला भारत विश्व का तीसरा देश है। दो अन्य देशों आस्ट्रेलिया एवं न्यूजीलैण्ड में इस प्रकार की अदालतें हैं। यहां इस न्यायाधिकरण से जुड़ी महत्वपूर्ण बातों का बिन्दुवार विवरण निम्नानुसार है:

- ☆ यह न्यायाधिकरण उन पर्यावरण संबंधी मामलों का निपटारा करेगा, जो देश की विभिन्न अदालतों में विचाराधीन है। प्रदूषित पर्यावरण से बीमारी, विकलांगता मृत्यु होने, व्यवसाय या सम्पत्ति को क्षति पहुंचने पर कोई भी व्यक्ति इस न्यायाधिकरण में मुआवजे का दावा कर सकेगा। पर्यावरण संबंधी हादसे के कारण जानमाल के नुकसान के एवज में फास्ट ट्रेक प्रणाली के तहत भी मुआवजे का दावा किया जा सकेगा।
- ☆ न्यायाधिकरण ऐसे प्रकरणों में पीड़ित पक्षकार की सहायता एवं क्षतिपूर्ति राशि (मुआवजा) प्रदान करने का आदेश जारी करने के लिऐ अधिकृत होगा। इस न्यायाधिकरण के निर्णय के विरूद्ध सर्वोच्च न्यायालय में ही अपील की जा सकेगी।
- ☆ प्राकृतिक संसाधनों के संरक्षण एवं पर्यावरण संबंधी वैधानिक अधिकारों से लैस इस राष्ट्रीय हरित न्यायाधिकरण की चार क्षेत्रीय बैंचे भी स्थापित की जाएंगी।
- ☆ पूरे देश में इस न्यायाधिकरण के अंतर्गत पर्यावरण न्यायालयों का एक ऐसा संजाल (Network) स्थापित किया जाएगा, जिसमें कि विषय विशेषज्ञ एवं दक्ष लोग सम्मिलित होंगे।
- ☆ पांच शाखाओं वाले इस न्यायाधिकरण में अधिकतम 20 पर्यावरण विशेषज्ञों एवं 20 न्यायाधीशों को रखे जाने का प्रावधान है।
- ☆ न सिर्फ देश का कोई भी नागरिक, अपितु कोई भी संस्था पर्यावरण से संबंधित मामला न्यायाधिकरण में दाखिल कर सकेगीं।
- ☆ सुनिश्चित प्रावधानों के तहत क्षति के लिए वह पक्षकार जिम्मेदार होगा, जो कि प्रदूषण फैलाने को दोषी पाया जाएगा।
- ☆ वैज्ञानिक शोधों एवं तर्को के आधार पर किसी कार्य को संपादित कराने वाले पक्षकार को यह साबित करना होगा कि उसके कार्य से पर्यावरण को किसी भी प्रकार की क्षति नहीं होगी।

पर्यावरण सुरक्षा एवं अंतर्राष्ट्रीय प्रयास

पर्यावरण का क्षरण एक वैश्विक समस्या है क्योंकि वायु, जल देश की सीमा में नहीं बंधते। समय–समय पर अंतर्राष्ट्रीय संधियों के द्वारा पर्यावरण की बेहतरी के प्रयास किए गए।

- ✰ अंटार्कटिका संधि (Antarctica Treaty)
- ✰ नाभिकीय मुक्त समुद्र तल संधि (Nuclear Free Seabed Treaty)
- ✰ जैविक हथियार अभिसमय (Biological Weapons Convention BWC)
- ✰ रासायनिक हथियार अभिसमय (Chemical Weapons Convention, CWC)
- ✰ स्टॉक होम संधि पैरिस संधि 2015

निष्कर्ष

कानून को लागू करने से जुड़े मुद्देः

पर्यावरण संरखण हेतु बनाए गए विभिन्न कानूनों को लागू करने में अनेक प्रकार की समस्याएँ सामने आती हैं। जैसे:

1. अशिक्षा व विधिक साक्षरता के अभाव में जनसंख्या का बहुत बड़ा भाग नहीं जानता है कि पर्यावरण के तत्वों जैसे :– जल, वायु, मृदा, वन आदि की गुणवत्ता में ह्यस करने पर कोई कानून भी लागू होता है।
2. छोटे उद्योग चलाने वाले या छोटी खदानों के मालिकों का आर्थिक पक्ष इतना सुदृढ़ नहीं होता कि अपशिष्टों का सुरक्षित निस्तारण कर सकें या ट्रीटमेंट प्लांट लगा सके।
3. वन्य जीवों व लकड़ी के तस्करों के अन्तर्राज्यीय गिरोहों का पकड़ने के मार्ग में विभिन्न राज्यों की सीमाऐं, अलग–अलग कानून, परस्पर तालमेल का अभाव, सीमित संसाधनों आदि बाधाओं के कारण सफलता नहीं मिल पाती है।
4. विभिन्न कानूनी प्रावधानों व खामियों का लाभ उठाकर अनेक बार उद्योगपति पर्यावरण कानूनों का उल्लंघन करके भी बच निकलते हैं कानूनी कार्यवाही को लंबित करने में सफल हो जाते हैं। संरक्षण के लिए 115 आर्द्रभूमि की पहचान की गई है। वर्ष 2012–13 में 36 आर्द्रभूमि प्रबंध कार्ययोजना मंजूर की गई तथा संबद्ध राज्य सरकारों को वित्तीय सहायता जारी की गई। राष्ट्रीय आर्द्रभूमि संरक्षण कार्यक्रम के अंतर्गत जारी तीन अनुसंधान परियोजनाओं के लिए भी वित्तीय किए गए।

पर्यावरण के मित्र की भांति जीने वाले मनुष्य की, कानून के बंधन से नियंत्रित करने की यात्रा में सामने आया निष्कर्ष मनुष्य की प्रवृत्ति को प्रदूषण के प्रमुख कारण के रूप में इंगित करता है।

प्रकृति की गोद मनुष्य की मॉ की भांति है, मॉ अपने बच्चे को दुलार देती है उसकी इच्छाएं पूरी करती है परन्तु जब बच्चे स्वयं को हानि पहुंचाने वाले कार्य करते हैं तो भी

नाराज होते है इसी तरह प्रकृति भी लगातार मानव समाज को देती रही है, प्रदूषण कार्य से राकेने के लिए भूकम्प के रूप में एशियन सुनामी 2004, फुकुशिमा दाईनी आपदा जापान, भूमिगत जल में फ्लोराइड प्रदूषण के हड्डियों पर पड़ने वाले प्रभाव, सूखा अम्लीय वर्षा ग्लोबल वार्मिंग है इनके अलावा–

चेरनोबिल – बड़ी मात्रा में रेडियोएक्टिव तत्वों का पर्यावरण में फैलाव, भोपाल – 45 टन विषेली मिथायल आइसोसायनेट से कुछ ही घंटों में हजारों लोगों की मृत्यु, कुवैत के तेल कुओं में आग, लैब केनाल (अमेरिका), एक्सन बॉलडेज, टोकाइमुरा परमाणु संयंत्र, अराल सी, सेवेसा डाइओस्नि क्लाउड मिनीमाटा डिसीज, थ्री माइल्स आइलैण्ड, अल मिश्राक बी. पी. तेल रिसाब, फुकुशिमा विकिरण।

उपरोक्त को नियंत्रित करने एवं सम्पोषी विकास के लिए मनुष्य को अपनी प्रवृत्ति में पूर्व भारतीय दर्शन आधारित जीवन शैली को अपनाना ही एकमात्र उपाय है।

वना पतये नमः वृक्षाणों पतये नमः

औषधींना पतये नमः अरण्यानां पतये नमः

यजुर्वेद की उक्त पंक्तियों में राष्ट्र की तरफ से वृक्षों, औषधियों एवं अरण्यों के रक्षक नियुक्त करने आर उन रक्षकों को उचित सम्मान देने का निर्देश मिलता है।

सुझाव

1. पर्यावरण संबंधी घटनाओं को अधिक से अधिक राष्ट्रीय हरित अधिकरण की दृष्टि से लाने का कर्त्तव्य हर नागरिक निर्वाह करे।
2. जनता के मध्य पर्यावरण प्रदूषण के कारण एवं निपटने की प्रक्रिया की जानकारी दूरदर्शन के माध्यम से।
3. अधिक से अधिक वृक्षारोपण – जन्म दिन आदि अवसरों पर।
4. शैक्षणिक संस्थाओं के ग्रीनबेल्ट अनिवार्य।
5. पर्यावरण के लिये कार्य करने वाले व्यक्त्यिों एवं कार्यों की जानकारी प्राथमिक स्तर पर पाठ्यक्रम में सम्मिलित किया जाना।
6. वन्य जीवन संरक्षण एवं प्रकृति के महत्व की डाक्यूमेंन्ट्री दूरदर्शन में प्रसारण एवं सिनेमाघरों में फिल्म के प्रारंभ में दिखाया जानां
7. जनसंख्या वृद्धि नियंत्रण हेतु योजना व प्रयास
8. गैर परम्परागत ऊर्जा स्त्रोतो को प्रोत्साहित किया जाना (सोलर एनर्जी जनित)
9. परिवार में बच्चों को पर्यावरण फ्रेन्डली दिनचर्या की आदत डलवाये एवं पर्यावरण के महत्व को समझाऐं।

10. हर राज्य अपने स्तर पर पुराने डीजल वाहनों को पुनः लाइसेन्स देने के स्थान पर नए वाहनों से प्रतिस्थापन की नीति बनाए।

11. हर राज्य में स्कूल में पुस्तकें प्रदाय कर अगले सत्र. के बच्चों के वितरण द्वारा बांस कटाई की बचत कर ।

गांधी जी ने अपने दूरदर्शी सोच में जो कहा था कि– "प्रकृति के पास मनुष्य की आवश्यकता पूरी करने की क्षमता है, मनुष्य की लालच पूरा करने की नहीं।" यह बात विश्व के मानव जितना जल्दी समझे उतना ही उनके स्वयं के अस्तित्व को बनाए रखने हेतु अच्छा है।

संदर्भ ग्रंथ

1. भारतीय संविधान ।
2. पर्यावरण संबंधित अधिनियम ।
3. अन्तर्राष्ट्रीय संधियां ।
4. यजुर्वेद ।
5. ए.आई..आर.

19

पर्यावरण संरक्षण और पोषणीय विकास में न्यायपालिका की भूमिका

भूपेन्द्र करवंद[1] एवमं श्रद्धा पांडे[2]

[1]विधि विभाग शास.जे.योगानंदम, छत्तीसगढ़ महा., रायपुर, छ.ग.

[1]विधि विभाग, मैट्स विश्वविद्यालय, रायपुर, छ.ग.

"दिल्ली में सांस लेना जहर हो गया है। छोटे–छोटे बच्चे यहां तक कि मेरा पोता भी मास्क पहनता है। जब वो मास्क लगाता है तो कार्टून कैरेक्टर निंजा जैसा दिखता है। हम चाहते हैं मसले को अखबार पहले पन्ने पर छापे। आमतौर पर अखबारों को ऐसा नहीं कहते, लेकिन लोगों को शिक्षित करना जरूरी है, इसलिये छापे।"

एच.एल. दत्तू (सी.जे.आई.)

दिनांक 05 अक्टूबर 2015 को मुख्य न्यायाधीष ने 1985 में दाखिल पर्यावरणविद् एम. सी. मेहता की याचिका पर सुनवाई करते हुये ये कथन कहे तथा पर्यावरण प्रदूषण पर गहरी चिंता व्यक्त की।

भारतीय न्यायालय ने विधियों का निर्वचन इस ढंग से किया है कि जिससे न केवल पर्यावरण संरक्षण हो बल्कि पोषणीय विकास का लक्ष्य भी प्राप्त हो जाये। वस्तुतः न्यायालय ने भारत में "पर्यावरणीय विधिशास्त्र" का प्रतिपादन किया है।

21वीं शताब्दी में न तो यह साध्य न व्यवहारिक है कि देश या समाज की विकास की तरफ नकारात्मक दृष्टिकोण अपनाया जाये परंतु इसका मतलब यह नहीं है कि पर्यावरण की तरफ ध्यान ही न दिया जाये। समाज को समृद्ध होना चाहिये परंतु पर्यावरण की कीमत पर

नहीं और इसी तरह पर्यावरण का संरक्षण होना चाहिये परंतु समाज के विकास की कीमत पर नहीं। अनुपोषणीय विकास ही एक मात्र साधन है।

पोषणीय विकास की संकल्पना मानव पर्यावरण का घोषणा सिद्धांत 2 से उत्पन्न होती है–''सतर्कतापूर्ण आयोजन और प्रबंध द्वारा पृथ्वी के प्राकृतिक स्त्रोतों जिनमें, हवा, पानी, पेड़–पौधे, जीव–जंतु, विशेष रूप से पारिस्थितिक तंत्र के नमूने सम्मिलित हैं, को वर्तमान और भावी पीढ़ियों हेतु सुरक्षित रखा जाना आवश्यक है।''

अर्थात् पोषणीय विकास वह विकास है जो वर्तमान आवश्यकताओं की पूर्ति भावी पीढ़ी की आवश्यकताओं की पूर्ति की सक्षमता के साथ बिना समझौता किये कर सके।

पोषणीय विकास की संकल्पना मानव पर्यावरण का घोषणा सिद्धांत 2 से उत्पन्न होती है–''सतर्कतापूर्ण आयोजन और प्रबंध द्वारा पृथ्वी के प्राकृतिक स्त्रोतों जिनमें, हवा, पानी, पेड़–पौधे, जीव–जंतु, विशेष रूप से पारिस्थितिक तंत्र के नमूने सम्मिलित हैं, को वर्तमान और भावी पीढ़ियों हेतु सुरक्षित रखा जाना आवश्यक है।''

अर्थात् पोषणीय विकास वह विकास है जो वर्तमान आवश्यकताओं की पूर्ति भावी पीढ़ी की आवश्यकताओं की पूर्ति की सक्षमता के साथ बिना समझौता किये कर सके।

इस संदर्भ में यह बता देनाप सुसंगत है कि किसी घोषणा का सिद्धांत 10 न्यायिक और प्रशासनिक प्रक्रियाओं में प्रभावकारी पहुॅच हेतु उपबंध करता है जिसमें प्रतितोष और उपचार सम्मिलित है।

भारतीय न्यायालय ने पर्यावरण संरक्षम में महत्वपूर्ण भूमिका अदा किया है और मामलों के निपटाने में पोषणीय विकास के सिद्धांत को लागू किया है।

रूरल लिटिगेशन एण्ड इण्टाइटिलमेण्ट केंन्द्र देहरादून बनाम उत्तर प्रदेश राज्य[1]

इस मामले में सन् 1983 में सरल लिटिगेशन एण्ड इण्टाइटिलमेंट केंद्र देहरादून नाम की एक स्वयं सेवी संस्था ने उच्चतम न्यायालय को पत्र लिखकर मसूरी के आसपास हिमालय की तलहटी में अवैधानिक तरीक से चलाये जा रहे चूना पत्थर खनन के कारण पर्यावरण पर पड़ने वाले विध्वंसकारी प्रभाव की ओर ध्यान आकृष्ट किया ।

उच्चतम न्यायालय के मुख्य न्यायाधीश पी.एन. भगवती, न्यायाधीश अमरेन्द्र नाथ और न्यायाधीश रंगनाथ मिश्रा की न्यायपीठ ने विस्तृत पट्टाधारक इस मामले में अन्तर्वलिन थे, इसलिये व्यापक सुनवाई करनी पड़ी। न्यायालय ने स्पष्ट किया कि यह देश में प्रथम मामला है जिसमें पर्यावरण और पारिस्थितिकी के मुद्दे उठे हैं।

---

1 AIR 1995 S.C. 652.

इस मामले में न्यायालय ने निम्न सिद्धांत प्रतिपादित किये:

1. पर्यावरण और पारिस्थितिकीय संतुलन की कीमत पर किसी तरह का व्यवसाय करने की अनुमति नहीं दी जा सकती ।
2. जहॉ विकास और पर्यावरण तथा पारिस्थितिकीय के बीच संघर्ष हो वहां दोनों के बीच सामंजस्य स्थापित किया जाना चाहिए ।

सोसायटी फॉर प्रोटेक्शन ऑफ साइलेंट वैली बनाम भारत संघ[2]

इस मामले में रिट याचिका द्वारा केरल राज्य के विरूद्ध व्यादेश की मांग की गई जिसमें राज्य शांति घाटी में जल विद्युत परियोजना का निर्माण न कर सकें । न्यायालय ने कहा:

''क्या राष्ट्रीय संपदा को जल विद्युत परियोजना के साथ ज्यादा उर्जा उत्पादन की संभावना को ज्यादा सुविधा पूर्वक प्रयोग में लाया जा सकता है, अथवा वनों व वन्य जीवों को इनकी प्राचीन गरिमा के साथ कायम रखने, भूमि रक्षण निवारण और समुदाय पर अन्य घातक प्रभावों को दूर करने में ज्यादा सुविधा पूर्वक प्रयोग में लाया जा सकता है। हम अपने निर्णय को सरकार के निर्णय के लिए स्थानापन्न नहीं कर सकते।''

इंडियन काउंसिल फॉर इनवायरो लिगल एक्शन बनाम भारत संघ[3]

इस मामले में ''प्रदूषक भुगतान करे'' के सिद्धांत की व्याख्या करते हुए सर्वोच्च न्यायालय ने कहा कि इस सिद्धांत की मांग यह है कि प्रदूषक को प्रदूषण से प्रभावित व्यक्तियों की क्षतिपूर्ति के साथ ही साथ पर्यावरण क्षति को पुनर्स्थापित करने की लागत को भी वहन करना होगा।

वेल्लोर सिटिजन वेलफेयर फोरम बनाम भारत संघ[4]

उच्चतम न्यायालय ने पूर्व सावधानी के सिद्धांत के अर्थ को स्पष्ट करते हुए निम्न बातों को सम्मिलित माना और कहा कि:

1. राज्य सरकार और स्थानीय प्राधिकारी द्वारा पर्यावरणीय उपाय पर्यावरण प्रदूषण के कारणों के बारे में पूर्वधारण, निवारण और समाधान किया जाये।
2. जहां पर्यावरण प्रदूषण से गंभीर और अपूर्णीय क्षति का खतरा है, वहां वैज्ञानिक निश्चितता के अभाव को निवारण उपायों के स्थगित करने का कारण नहीं बनाया जावेगा।

---

2 OP Numbers, 2499 & 3025 of 1979 (unreported).

3 (1996)3 S.C.C. 212.

4 (1996) 5 S.C.C. 647.

3. पर्यावरणीय रूप से कार्यवाही के अहितकर न होने को साबित करने का भाग प्रयोगकर्ता, विकासकर्ता और उद्योगपति पर होगा।

एम.सी. मेहता बनाम कमलनाथ व अन्य[5]

इस मामले में माननीय उच्चतम न्यायालय ने लोक न्यास के सिद्धांत को भारत में लागू किया। प्रस्तुत मामले में कमलनाथ सन् 1990 में पर्यावरण और वनमंत्री रहते हुए व्यास नदी के निकट एक रिसार्ट का निर्माण कराते हुए नदी की दिशा को मोड़कर 27.12 बीघा जमीन अतिक्रमण कर ली, जिससे प्राकृतिक नदी में छेड़छाड़ के कारण पर्यावरण संतुलन बिगड़ गया और बाढ़ के प्रकोप का सामना करना पड़ा। इसमें एम. सी. मेहता की एक जनहित याचिका में माननीय न्यायालय ने संज्ञान लेते हुए कमलनाथ व अन्य पर 105 करोड़ का जुर्माना अधिरोपित करते हुए व्यास नदी को पुनः उसके प्राकृति स्वरूप में लाने हेतु निर्देश दिये, और सरकार को सख्ती से कहा कि सभी प्राकृतिक संपदाओं को उसकी प्राकृतिक विशेषताओं के साथ बनाये रखना सरकार का कर्तव्य है। क्योंकि ये संपदाएँ सरकार के पास लोक न्यास में होती हैं और सरकार केवल उसका न्यासी होता है।

एम. सी. मेहता बनाम भारत संघ[6]

इस वाद में उच्चतम न्यायालय ने परिसंकटमय पदार्थों और स्वभावतः घातक पदार्थो संबंधी उद्योगों को दूसरे स्थान पर अंतरित करने के आदेश को पारित किया। वहीं न्यायालय ने यह भी मत व्यक्त किया कि इन उद्योगों को पुनः चालू करने के पूर्व यह सुनिश्चित कर लिया जाये कि वे जल प्रदूषण एवं वायु प्रदूषण की शर्तो को पूरा करते हैं । इस हेतु न्यायालय ने एक विशेषज्ञ समिति का गठन किया जो संयंत्र में विद्यमान खामियों को दूर करने के उपाय बनाएं।

मुरली देवरा बनाम भारत संघ

स्वास्थ्य के प्रति जागरूकता की आवश्यकता को देखते हुए उच्चतम न्यायालय ने सार्वजनिक स्थानों पर धुम्रपान के कुप्रभाव की ओर सरकार का ध्यान आकर्षित कराते हुए निर्देश जारी किया कि वे सार्वजनिक स्थानों पर धुम्रपान को निषिद्ध करने हेतु प्रभावकारी कदम उठाए।

निष्कर्ष

भारत में पर्यावरण के संरक्षण और उसके सुधार करने के लिए संसद और राज्य विधान मंडल ने भारी भरकम संख्या में अधिनियमों को पारित किया है और उसके आधार पर

---

5 (1997) 1 S.C.C. 388.

6 AIR 1987 S.C. 965.

प्रशासनिक प्राधिकारियों ने विशालकाय संख्या में नियमों और नियमावलियों को बनाया है। पर्यावरण के संरक्षण और सुधार के मामले में न्यायपालिका भी पीछे नहीं है और इस संबंध में माननीय उच्च न्यायालय और उच्चतम न्यायालय ने पर्यावरणीय विधि शास्त्र का सृजन किया है और न्यायिक मान और मानकों का निरूपण और प्रतिपादन किया है।

संदर्भ

1. डॉ. उपाध्याय, जय जय राम – पर्यावरण विधि.
2. डॉ. पाण्डेय, जयनारायण – भारत का संविधान.
3. प्रो. चंद्रा, सतीश – भारत का संविधान–एक समीक्षा.
4. डॉ. तिवारी, ए.के. – भारत में पर्यावरण विधियॉं.
5. प्रो. सिंग, गुरूदीप – भारत में पर्यावरण विधि.

20

वर्तमान साहित्य में पर्यावरण विमर्श

अनसूया अग्रवाल

हिन्दी विभाग,
शासकीय महा. वल्लभाचार्य स्ना. महा. महासमुन्द (छ.ग.)

वर्तमान साहित्य का संघर्ष कई तरह के विमर्श को समाहित किए हुए है जिसमें नारी, दलित, व्यवस्था, भाषा के अतिरिक्त पर्यावरण भी कविता की केन्द्रीय चिन्ता का एक उल्लेखनीय पक्ष है। हमारी संवेदनहीनता, यथार्थ से विमुखता, तथा भौतिक संपन्नता की भूख ने ऐसी भयानक स्थिति उत्पन्न कर दी है कि हमारी जानी–पहचानी प्रकृति, फूल–पत्ते, वनस्पतियॉ आदि सब स्वप्न गाथा बनकर रह गए हैं। मिट्‌टी से हमारा रिश्ता सतही हो गया है। हमॉरे कवियों को लगता है कि आज की अनेक समस्याएँ प्राकृतिक असंतुलन के कारण हैं। ओजोन परत में छेद हो जाना कोई मामूली समस्या नहीं है क्योंकि इसके दुष्परिण ााम दूरगामी है। कवियों की चिंता इन्हीं बातों को लेकर है कि जब तकनीकी सभ्यता धरती के स्नायु तंत्र को छिन्न–भिन्न करेगी, प्रकृति की स्वाभाविकता को नष्ट करेगी तो प्रकृति तो विक्षुब्ध होगी ही न? कवि शिशुपाल सिंह की भूकंप पर आधारित 'नहीं चाहिए मुझे एक और लातूर' शीर्षक कविता में प्रदूषण के रोकथाम की अपील की गई है:

> ''फैल रहा है प्रदूषण विकृति और विक्षोभ, यथास्थिति से कहॉ मुक्ति
>
> कब तक चलेगा यह निर्मम चक, नहीं चाहिए मुझे एक और लातूर।''

कुछ ही दिन हुए वर्ष 2016 के आगमन को; यानि इक्कीसवीं सदी को आये पन्द्रह बरस बीत चुके हैं किंतु गंगा, यमुना, सरस्वती आज भी तो मैली हो रही हैं। कमलेश भट्‌ट 'कमल' नदी के प्रदूषण के प्रश्न को बड़ी मार्मिकता से उभारते हैं। नदी का क्षत– विक्षत होना कवि को बहुत त्रास देता है। कवि का भयभीत मन कह उठता है:

"हमने वर्षो विष पिला कर आजमाया है बहुत,
अब हमें भी विष पिलाकर आजमाएगी नदी।"

अस्वछता, प्रदूषण, ग्लोबल वार्मिंग, भूकंप, बाढ़, अकाल, सूखा, अनावृष्टि, घटता हुआ जलस्तर असमय मौसम परिवर्तन, समुद्री तूफान पहले की तरह ही बढ़ रहे हैं और लगता है आने वाले दिनों में प्राकृतिक विभीषिकाएं और भी गहराएँगी। हम भौतिक रूप से जितने सभ्य और संपन्न होते जा रहे हैं, सांस्कृतिक और पर्यावरणिक रूप से उतने ही असभ्य और निर्धन होते जा रहे हैं। ये सब प्रकृति के प्रति अप्राकृतिक आचरण का कुफल है। कन्हैया लाल 'वक्र' अपनी कविता 'पहाड़' में अत्यंत मार्मिक ढ़ंग से कहते हैं:

"और अब पेड़ कहाँ हैं?

ये तो कुछ खूँटे गड़े हैं,

पेड़ सारे आदमी के पेट में पड़े हैं।"

विनोद पदरज पेड़ों को अपनी विरासत से जोड़ते हुए अपनी कविता 'पिता और पेड़' के माध्यम से हमें सावधान करते हैं यह कहकर कि 'पेड़ के कटने का मतलब अपनी समूची परंपरा पर आघात है':

"खचाक् खचाक्

काटा जा रहा है पेड़,

मेरी नींद लहू– लुहान हो रही है।"

कवियों की पीड़ा और दर्द उनके काव्यों में तब शिद्दत से उभरता है जब वे देखते हैं कि औद्योगिक और नगरीकरण की अंधी दौड़ में वह प्रकृति क्षत– विक्षत हो रही है जिसके कोड़ में मानव पला और बढ़ा है। उन्हें यह स्वीकार्य नहीं कि विकास जीवन और पर्यावरण की कीमत पर किया जाए क्योंकि यदि संस्कृति हमारे जातीय गौरव को बढ़ाती है तो पर्यावरण हमारे जैविक गौरव को। "सौ–सौ नदियों क जल वाला, ऊँचे–ऊँचे पर्वत माला पर्यावरण का सौंदर्य भी हमें इसकी रक्षा में आत्मोसर्ग करने की प्रेरणा प्रदान करता रहा है। पर्यावरण के बिना हम एक पल भी सॉस नहीं ले सकते।" (1) प्रगतिशील कवि नंद चतुर्वेदी का संवेदनशील मन तब व्यथित हो उठता है जब वो देखते हैं कि मनुष्य का अपनी मिट्टी या जमीन से रिश्ता टूटने लगा है:

"अपनी आत्मा के दुःख बहुत से हैं

सबसे बड़ा दुःख तो यह है कि

वह वट–वृक्ष जिसकी टहनियॉ

कितने नीचे ठीक जमीन को छूती थीं

अब सिर्फ हाथ भर की रह गयी हैं

जमीन के रिश्ते, जमीन, मिट्टी, पानी

ये फूल पत्ते

एक स्वप्न गाथा गाने लगे हैं।''

कवि जानता है कि कटते हुए पेड़ों के साथ आदमी भी मरता है क्योंकि वृक्षों, पहाड़ों, नदियों से हमारा रिश्ता जमीनी नही है; इनका होना घर के वयोवृद्व का होना है। 'सुनो कारीगरः पिता' कविता में उदयप्रकाश ने पर्यावरण संरक्षण का संदेश कुछ इसी अंदाज में दिया हैः

''पिता

झाड़– झंखड़, घाटियों और पत्थरों से भरे

चटियल मैदान थे

पिता

सागौन, शीशम, बबूल, तेंदुओं और हिरनों से भरे

बीहड़ थे

पिता

पहाड़ की तरह

चलते भीड़ के बीच..............।''

डॉ. वेदप्रकाश अमिताभ लिखते हैं– ''न केवल जंगलों का कटाव अपितु अवैध उत्खनन, महानगरीय मलिनता का नदियों में उत्सर्जन, गैस रिसाव आदि हादसे भी कवियों को उद्व ेलित करते हैं। उनके माध्यम से नदियों की शिकायत हैः

न जाने कब पड़ गयी हम पर सभ्य आदमी की असभ्य नजर

हमारी छाती पर खोल दिए इसने

तेजाबी गटर।'' (2)

दरअसल भौतिकवाद और पूंजीवाद के चलते हम भी शोषक बन बैठे हैं। जंगलों– पहाड़ों को काटना, धरती की कोख से खनिज निकालना, जीव– जन्तुओं का भक्षण करना, नदियों को पाटकर मैदान बनाना आदि शोषण के ही अंग है। अपने स्वार्थ के खातिर, अपने धन को जुटाने– बढ़ाने के खातिर हम पर्यावरण का जितना संभव हो; शोषण कर रहे हैं। इसके पीछे हमारी मूल भावना सब कुछ पा लेने की है। बाद में कोई दूसरा न पा सके। भले ही इस कोशिश में मौजूदा भी खत्म क्यों न हो जाए। ''यूज एण्ड थ्रो'' की हमारी प्रवृति भीतर

के आदमी को नष्ट कर रहा है। अब मानवीय खूबियों से हम रिक्त होने लगे हैं। दया, प्रेम, करूणा, सहानुभूति अपनत्व की भावनाएँ सूखती जा रही हैं। पूंजीवादी मानसिकता, बाजारवाद आदि ने एक अमानवीयकरण को जन्म दिया है। मनुष्य अपनों के प्रति और प्रकृति के प्रति बेरहम हो चला है। रामदरश मिश्र की कविता इस त्रासदी को बखूबी विस्तार देती है:

"उसने सोने– चॉदी के पहाड़ पर

बारूद के पेड़ लगाए हैं

जिनकी विषाक्त सॉसों से

झुलस गयी हैं वनस्पतियॉ

टूट गए हैं चिड़ियों के पंख

तेजाब सा खलबला रहा है नदियों का पानी

जल गए मौसमों के रंग

और वे एक– दूसरे में समा गए हैं

और आदमी

एक प्यासा शोर बनकर रह गया है। " (3)

लगभग सभी कवियों का प्रकृति के साथ गहन लगाव है। सबकी कोशिश अपनी धरती को बचाने की है। ओमप्रकाश मेहरा सर्वथा सार्थक और सकारात्मक अपील करते हुए लिखते हैं:

गगनभेदी तोपें उगलती हैं

जहरीले धुँए।

आओ इन धुओं से हम

अपने निर्मल आकाश को गॅदला होने से बचाएँ

निसंदेह आज आवश्यकता प्राकृतिक संतुलन को बनाए रखने की है। डॉ. वंदना श्रीवास्तव लिखती हैं– " अब समय कम है और काम का विस्तार बड़ा है। यदि हम आज नहीं जागे तो वह दिन दूर नहीं जब हमारी पृथ्वी पर जीवन समाप्त हो जाएगा। इसलिए हमें यह मंत्र अपनाना होगा–आज, अभी, अभी नहीं तो कभी नहीं।" (4)

निष्कर्षतः कह सकते हैं कि विकास के नाम पर प्रकृति से खिलवाड़ उचित नहीं। प्रकृति भी एक सीमा तक ही हमारी ज्यादती को सहन कर सकती है। यदि प्रकृति अनाचार को नहीं सह पाई तो इसका परिणाम संपूर्ण मानव समाज के लिए अत्यंत घातक होगा अतः हमें ग्रेस कुजूर की इस चेतावनी का सदैव स्मरण रखना होगा:

"न छेड़ो प्रकृति को

अन्यथा

एक दिन

माँगेगी

हमसे

तुमसे

अपनी तरूणाई का

एक–एक क्षण

और करेगी

भयंकर बगावत।"

संदर्भ ग्रंथ

1. मधुमतीः दिसम्बर 2006, पृ. 35, सं.– डॉ. श्रीमती अजित गुप्ता, राजस्थान साहित्य अकादमी, उदयपुर में प्रकाशित, 'साहित्य के परिप्रेक्ष्य संस्कृति और पर्यावरण'– डॉ. ममता कालिया के आलेख से।
2. मधुमतीः फरवरी 2005, पृ. 18, सं.– अभय कुमार, राजस्थान साहित्य अकादमी, उदयपुर में प्रकाशित, 'समकालीन काव्यः पर्यावरण' विमर्श'– डॉ.वेदप्रकाश अमिताभ के आलेख से।
3. रामदरश मिश्र रचनावली खंड– 2, पृ. 163।
4. कालजयी, प्रवेशांक मार्च 2015, पृ. 38, प्रधान संपादकः डॉ. लारी आजाद, दोधपुर, अलीगढ़, उत्तर प्रदेश से प्रकाशित, 'अभी नही ंतो कभी नहीं'– डॉ. वंदना श्रीवास्तव के आलेख से।

21

पर्यावरण प्रदूषण, स्वास्थ्य और प्रबंधन

मधुबाला मिश्रा एवमं रीता यादव

हिन्दी विभाग, शास. राजीव लोचन महावि. राजिम

सार-संक्षेप

प्राचीन काल में मानव, ऋषिमुनि कंदराओं गुफाओं तथा प्रकृति के सहज सुरम्य वातावरण में निवास किया करते थे। पेड़, पौधे, जल, वायु, मिट्टी उनके लिए देवी देवता के सूचक थे। वे उनके संरक्षक थे। पर आज वातावरण में आक्सीजन की कमी, विषैले गैसों की अधिकता है, जो हमारे भूल की ओर संकेत करती है। विकास की दौड़ में हम पर्यावरण का दुरूपयोग कर रहे हैं, जिसका दुष्परिणाम सामने है। पेड़ पौधे की कटाई से रेगिस्तान, सूखा, भूमिगत जल में कमी हो रही है। उद्योगों, कारखानो की चिमनी के धुएँ हमारी औद्योगिक संपन्नता को दर्शाते है, पर प्रश्न उठता है क्या हम प्रगति कर रहे है? वाहनो, चिमनियो से निकले धुएँ का पूर्व उपचार नही किया जाता फलस्वरूप हम विभिन्न रोगो से ग्रसित हो रहे है। जल जीवन के लिए अत्यन्त आवश्यक है। ज्ञान के अभाव में हम नदियों, नालो में कूड़ा, कर्कट तथा उद्योगों से निकले उत्सर्जित, रासायनिक पदार्थ खुला छोड़ देते है, जिससे जल प्रदूषण की गंभीर समस्या उत्पन हो गई है। प्रदूशित जल के उपयोग से बीमारियों में पीड़ित हो रहे है। कारखानों वाहनों की तीव्र ध्वनि प्रदूषण पढ़ गया है। मिट्टी भी प्रदूशित हो रही है, लगातार रायसानिक उर्वरक के उपयोग से जो मानव के स्वास्थ्य के लिए हानिकारक है। मनुष्य की आयु अल्प हो रही है, मृत्यु दर बढ़ रही है।

ओजोन परत का क्षरण, धरती का तापमान बढ़ना, जलवायु में परिवर्तन होने से पूरा विश्व प्रभावित होगा। अनेक संक्रामक रोग, उत्पादन में कमी, जलीय जंतुओं के प्रजनन आदि पर पर्यावरण के दूषित होने का प्रभाव होगा। पर्यावरण की अवहेलना कर मानव विकास नही कर सकता। पर्यावरण सम्मेलनों, विभिन्न सर्वेक्षणों तथा विश्व स्वास्थ्य संगठन ने विकसित

एवं विकासशील देशों को आगाह किया है कि पर्यावरण को संतुलित रखने के लिए उन्हे हर संभव उपाय अपनाया जाना चाहिए। पर्यावरण जागरूकता, पेड़ पौधे लगाकर, तीव्र ध्वनि पर पाबंदी, नदी, नालो की स्वच्छता के लिए कड़े नियम बनाकर पर्यावरण को स्व्च्छ रखा जा सकता है, और मानव को विनाशसे बचाया जा सकता है।

संकेत शब्दः पर्यावरण प्रदूषण, स्वास्थ्य, प्रबंधन.

प्रस्तावना

मानव के चारों ओर का वह क्षेत्र जो उसे घेरे रहता है, उसके जीवन तथा क्रियाओं को प्रयत्क्ष या अप्रत्यक्ष रूप से प्रभावित करता है पर्यावरण कहलाता है।(1) मानव के विकास में पर्यावरण का अत्यधिक महत्व है। पर्यावरण में हम एक दूसरे से जुड़े हुए हैं। पर्यावरण में असंतुलनकी स्थिति से सभी को खतरा का सामना करना पड़ेगा। विकास की प्रक्रिया में मनुष्य एक ओर उन्नति के शिखर चूम रहा है, तो दूसरी ओर विभिन्न समस्याएँ उत्पन्न हो रही हैं। पर्यावरण में तेजी से बदलाव खतरनाक स्थिति की सूचना है। धरती का तापमान बढ़ना जलवायु परिवर्तन, सूखा, अल्पवर्षा, अतिवर्षा, असमय वर्षा, इन सभी प्राकृतिक प्रकौप से मनुष्य भयभीत हो रहा है। आखिर क्यो ?

प्रस्तुत शोघ पत्र में पर्यावरण प्रदूषण के स्त्रोत, प्रदूषण का स्वास्थ्य पर प्रभाव तथा पर्यावरा प्रबंधन पर विश्लेषण किया गया हैः

अध्ययन का उद्देश्य

प्रस्तुत शोध पत्र के निम्न उद्देश्य निर्धारित किए गए हैः

1. पर्यावरण प्रदूषण के स्त्रोत का अध्ययन करना।
2. पर्यावरण प्रदूषण का स्वास्थ्य पर प्रभाव का अध्ययन करना।
3. पर्यावरण प्रदूषण के नियंत्रण के उपायों का अध्ययन करना।

अध्ययन क्षेत्र (सामान्य जन जीवन)

सामान्य जन जीवन की ओर दृष्टि डालते है तो स्पष्ट होता है कि हर व्यक्ति राष्ट्र विकास करना चाहता है। विकास की प्रक्रिया में वनों की कटाई से पेड़ पौधों की कमी हो रही है जिससे वातावरणमें आक्सीजन की कमी हो रही है। उद्योग धंधों, वाहनों, से निकलती विषैली धुएँ वातावरण को प्रभावित कर रही है।2 इससे वायु दुषित होकर स्वास्थ्य पर कु प्रभाव डाल रही है। ''जल'' जीवन के लिए अति आवश्यक है पर हम उसमें भी कचरा, उद्योगों का उत्सर्जित पदाथ, उपचारित किए बिना बहा देते है। एक ओर 3 प्रतिशत जल ही पृथ्वी पर मीठा है विमें 0.003 प्रतिशत नदी तालाब, में उपयोग हेतु उपलब्ध है3 और वह भी प्रदूषित हो तो क्या हो विचारणीय प्रश्न है। कल कारखानों उद्योगों की तेज ध्वनि शोर उत्पन्न करती ह। ध्वनि प्रदूषण भी समस्या की स्वास्थ्य पर कानों पर, प्रस्तुत शोध पत्र में पर्यावरा प्रदूषणों के कुछ स्वरूप 1. वायु प्रदूषण, 2. जल प्रदूषण, 3. ध्वनि प्रदूषण, का स्वास्थ्य

पर पड़ने वाले प्रभाव को निम्नकिंत तालिका 21.1–21.3, में दर्शाया गया है:

तालिका 21.1: वायु प्रदूषण का स्वास्थ्य पर प्रभाव

क्रं.	स्त्रोत	प्रदूषक	स्वास्थ्य पर प्रभाव
1.	वाहन	विभिन्न गैसे CO	सिर दर्द, बैचैनी
2.	उद्योग धंधे	धुएँ	हृदय रोग,दमा,छींक
3.	घरेलु उपयोगी चूल्हा, सिगड़ी	राख, धूल, (CO, NO, SO_2)	खॉसी, श्वांस, रूकना, दम घुटना आदि

तालिका 21.2: जल प्रदूषण का स्वास्थ्य पर प्रभाव

क्रं.	स्त्रोत	प्रदूषक	स्वास्थ्य पर प्रभाव
1.	उद्योगो द्वारा उत्सर्जित पदार्थ	उत्सर्जित पदार्थ K, Na, Ca, NH_3	डायरिया पीलिया, विकलांगता, कालरा
2.	कचरा, पालीथीन जलकुम्भी	CO_2 की अधिकता	कैंसर, त्वचारोग, आदि

तालिका 21.3: ध्वनि प्रदूषण का स्वास्थ्य पर प्रभाव

क्रं.	स्त्रोत	प्रदूषक	स्वास्थ्य पर प्रभाव
1.	मोटर वाहनों	हार्न, करकस, तेज ध्वनि	बहरापन,मानसिक असंतुलन
2.	कारखानों की मशीनों की आवाजे	शोर, अप्रिय ध्वनि	थकावट, चिड़चिड़ापन, कान का पर्दा फटना आदि

उपरोक्त तालिका 21.1–21.3 से स्पष्ट है कि "पर्यावरण की गुणवत्ता ह्रास चिंता का विषय है पर्यावरण मानव के प्रतिकूल है। 4 उपरोक्त तालिकाओं का अध्ययन करने पर निम्न तथ्य सामने आ रहे है:

1. पर्यावरण प्रदूषण की मानीटरींग में लापरवाही।
2. पर्यावरण की स्वच्छता पर लापरवाही।
3. पर्यावरण प्रदूषण आधिनियम के पालन में शिथिलता।
4. मनुष्यों की पर्यावरण के प्रति लापरवाही

पर्यावरण के संतुलन से ही विकास का मार्ग खुल सकता है। पर्यावरण प्रबंधन ही एक

मात्र मार्ग है, जिसके द्वारा संकट को नियंत्रण किया जा सकता है। पर्यावरा प्रबंधन के अंतर्गत नियोजन, विश्लेषण मूल्यांकन एवं उचित निर्णय द्वारा सीमित संसाधनों का उपयोग तथा प्राथमिकताओं में परिर्वतन आवश्यक है।

पर्यावरण प्रबंधन का एक महत्वपूर्ण पक्ष प्रदूषण नियंत्रण है जिन्हे कड़ाई से लागू किया जाना चाहिए एवं पर्यावरण प्रबंधन संबंधी सुझाव एवं उनके लाभ को निम्नानुसार तालिका 21.4 से समझा जा सकता है:

तालिका 21.4: पर्यावरण प्रबंधन संबंधी सुझाव

क्रं.	पर्यावरण प्रबंधन संबंधी सुझाव	लाभ
1.	पर्यावरण अधिनियमों का कड़ाई से पालन।	समस्त प्रदूशण (जल, ध्वनि, वायु) में कमी
2.	वक्षा रोपण	वातावरण में O_2 की अधिकता
3.	सायकल मित्र अपनाएँ	फुर्तीला स्वास्थ्य
4.	तुलसी, कनेर, मदार, बरगद के पेड़ लगाने को प्राथमिकता	प्रदूषण पोषक (स्वच्छ वातावारण)
5.	जल शुद्धिकरण संयत्र	बीमारियेां से राहत
6.	पर्यावरण जागरूकता संबंधी दिवस, कार्यक्रम को बढ़ावा	सामान्य जन जागरूकता का गुण आना

अंततः कहा जा सकता है, पर्यावरण को स्वच्छ रखकर ही हम विकास कर सकते है। पर्यावरण की अवहेलना कर हम जीवन की कल्पना नहीं कर सकते है। सरकार व प्रत्येक व्यक्ति को पर्यावरण के प्रति सचेत रहना होगा तभी देश खुशहाल होगा।

संदर्भ ग्रंथ

1. दीप्ती रानी, 2005, पर्यावरण अध्ययन, पृष्ट 1 आराधना प्रकाशन 206, जयपुर हाऊस आगरा।
2. दृष्टव्य दबंग दुनिया (समाचार पत्र)–2जनवरी 2016 शंकरलाल पुरोहित, रतलाम।
3. शुक्ला शशि, तिवारी ए.के.2010 पर्यावरण अध्ययन, पृ. 129 रामप्रसाद एंड संस अस्पताल रोड़ आगरा–3।
4. दृष्टव्य, गुर्जर आर के, जाट बी.सी., 2004, पर्यावरण अध्ययन, पृ. 148, पंचशील, प्रकाशन, फिल्म कालोनी, जयपुर.
5. शुक्ला शशि, तिवारी ए. के, 2010 पर्यावरण अध्ययन, पृ 11 रामप्रसाद एंड संस अस्पताल रोड–आगरा/3
6. दृष्टव्य मुखर्जी अंजली, 2010 पर्यावरण अध्ययन, पृ. 154–55, छ.ग.राज्य हिन्दी ग्रंथ अकादमी,

Index

T

V

W

Z

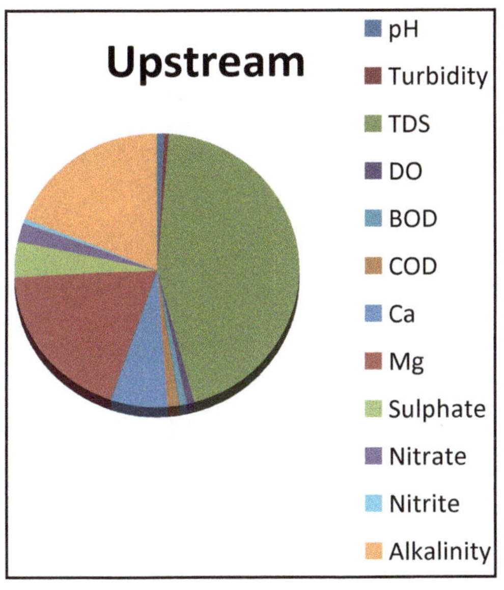

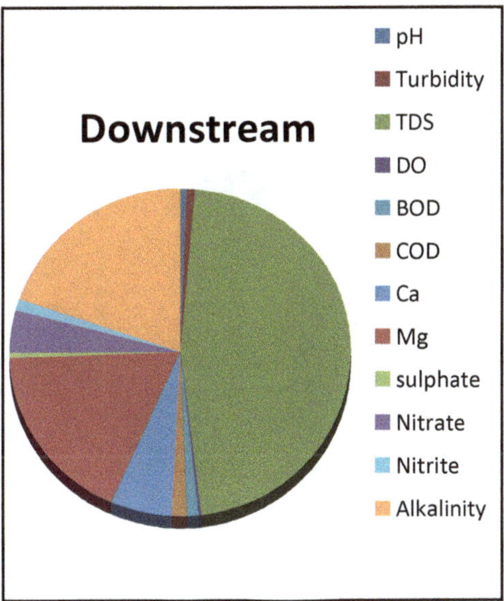

(Page 7)

1. PDA Plate with Malachite Green. (Page 13)

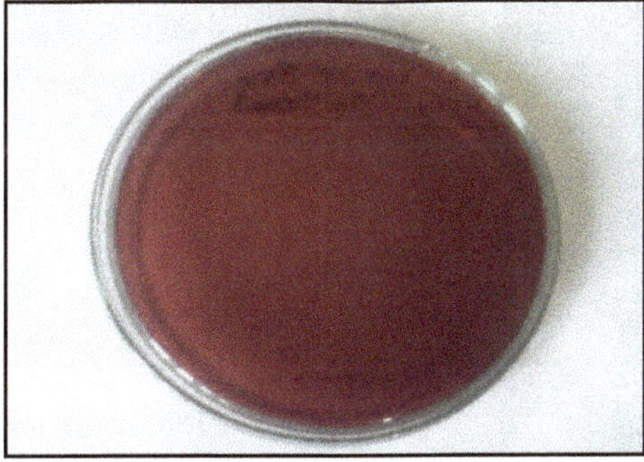

2. PDA Plate with Congo Red. (Page 13)

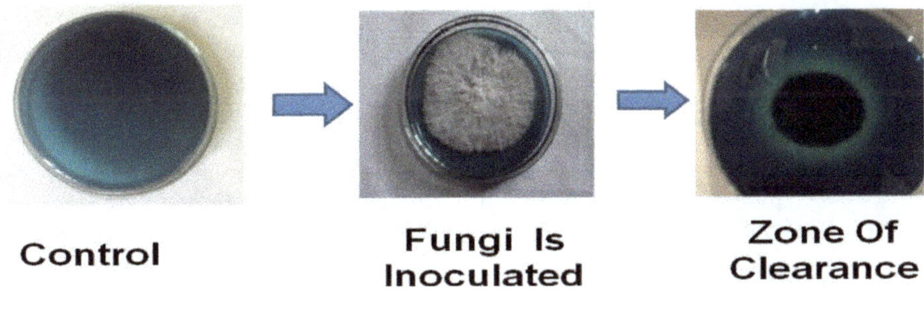

(Page 14)

After 5 Days **After 10 Days**

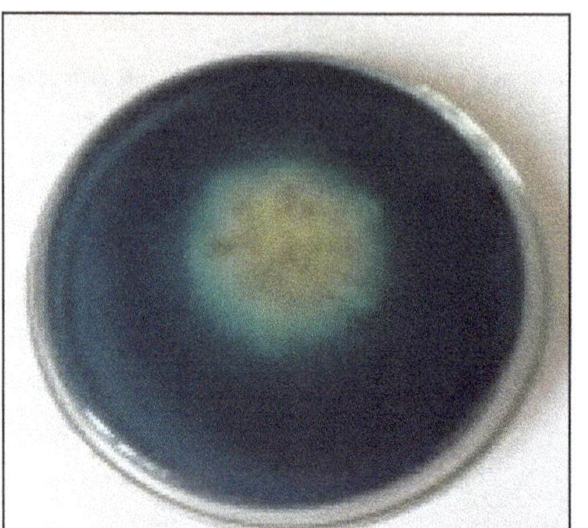
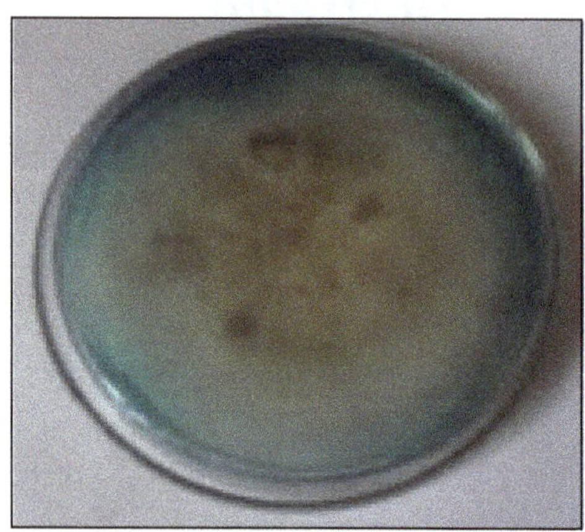

Pleurotus florida **(Page 14)**

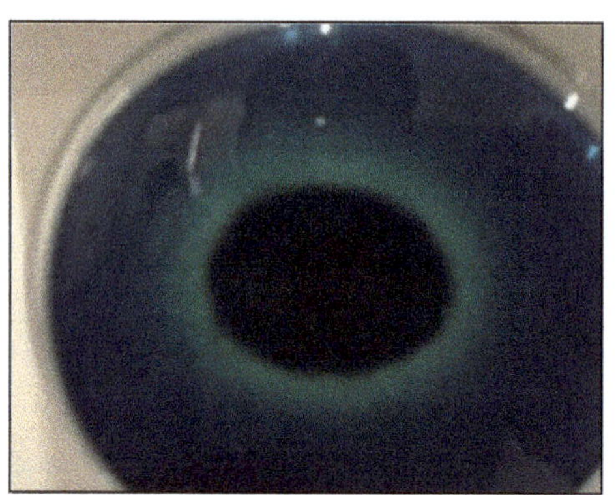
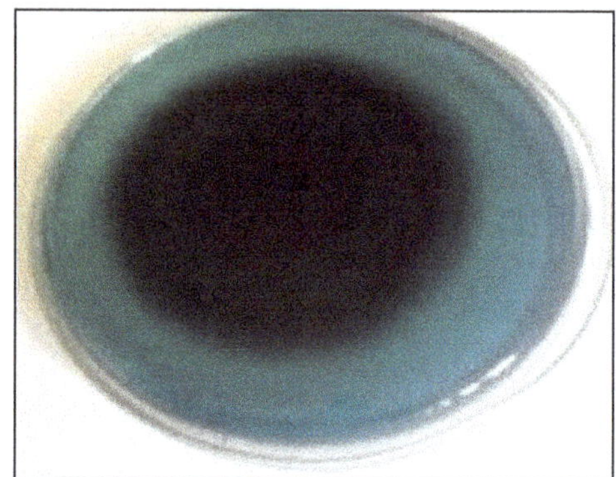

Pleurotus sajor-caju **(Page 15)**

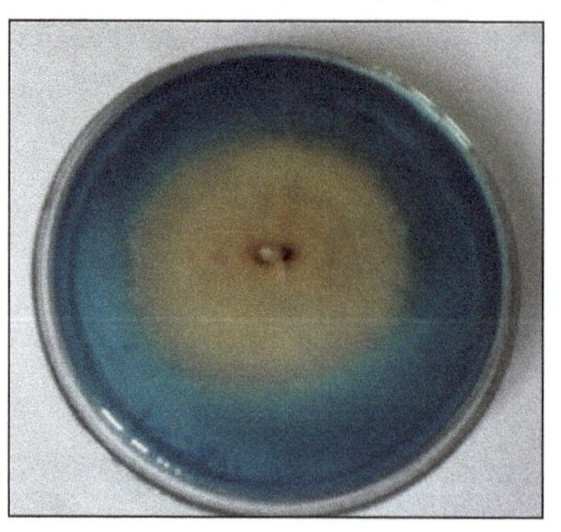
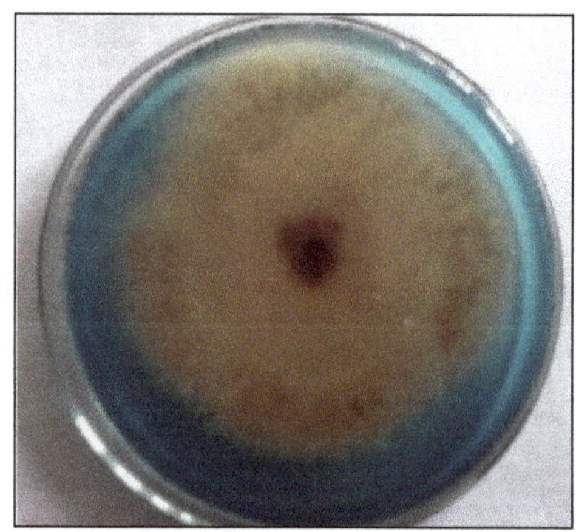

Polyporus **sp 1 *(Page 15)***

After 5 Days **After 10 Days**

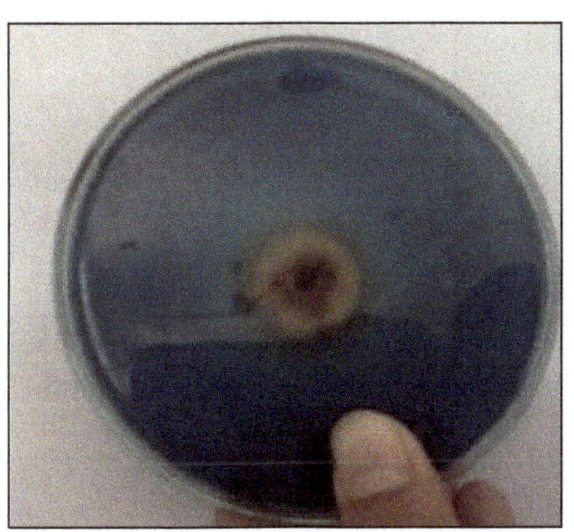

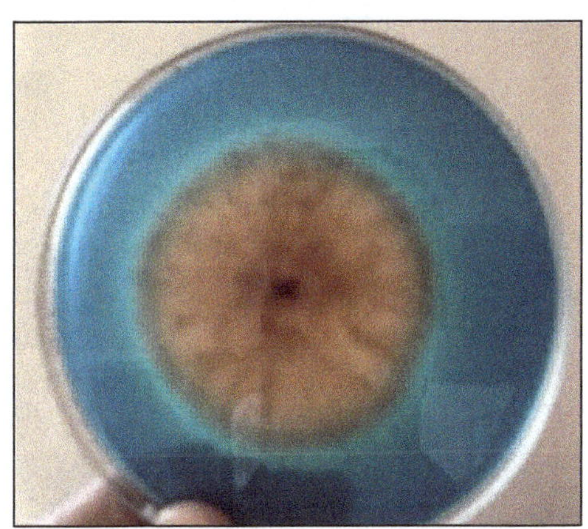

Schizophylum commune (Page 15)

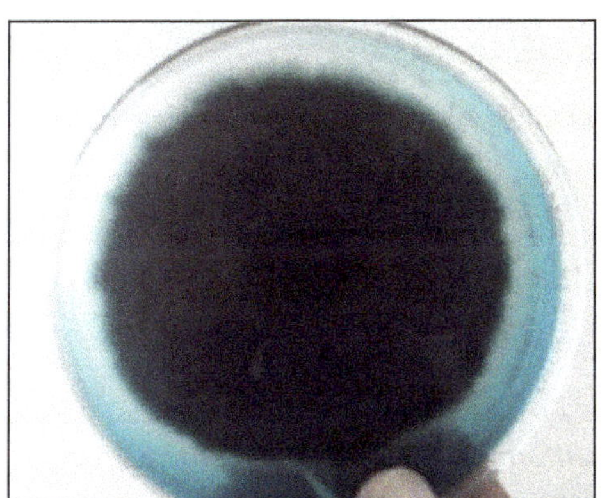

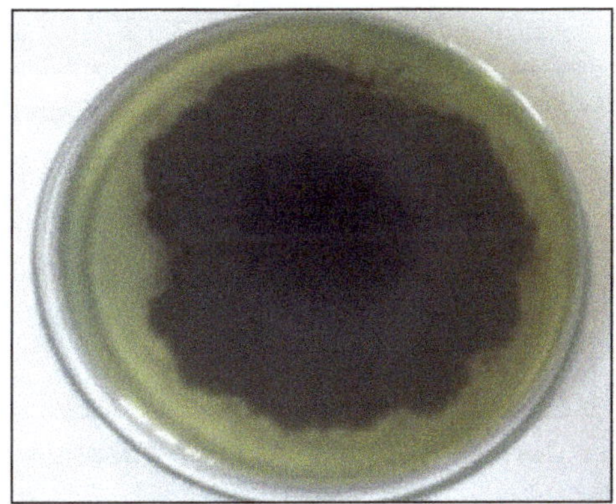

***Jelly* sp. (Page 16)**

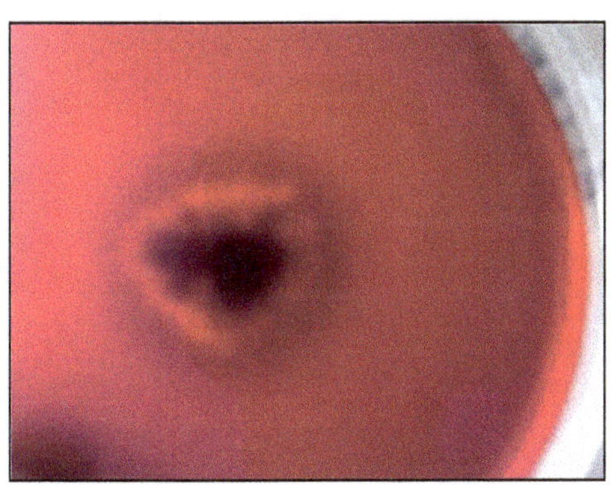

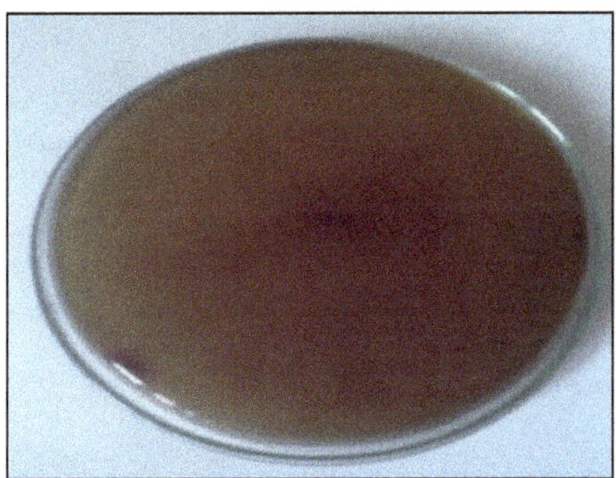

Pleurotus sajor-caju (Page 16)

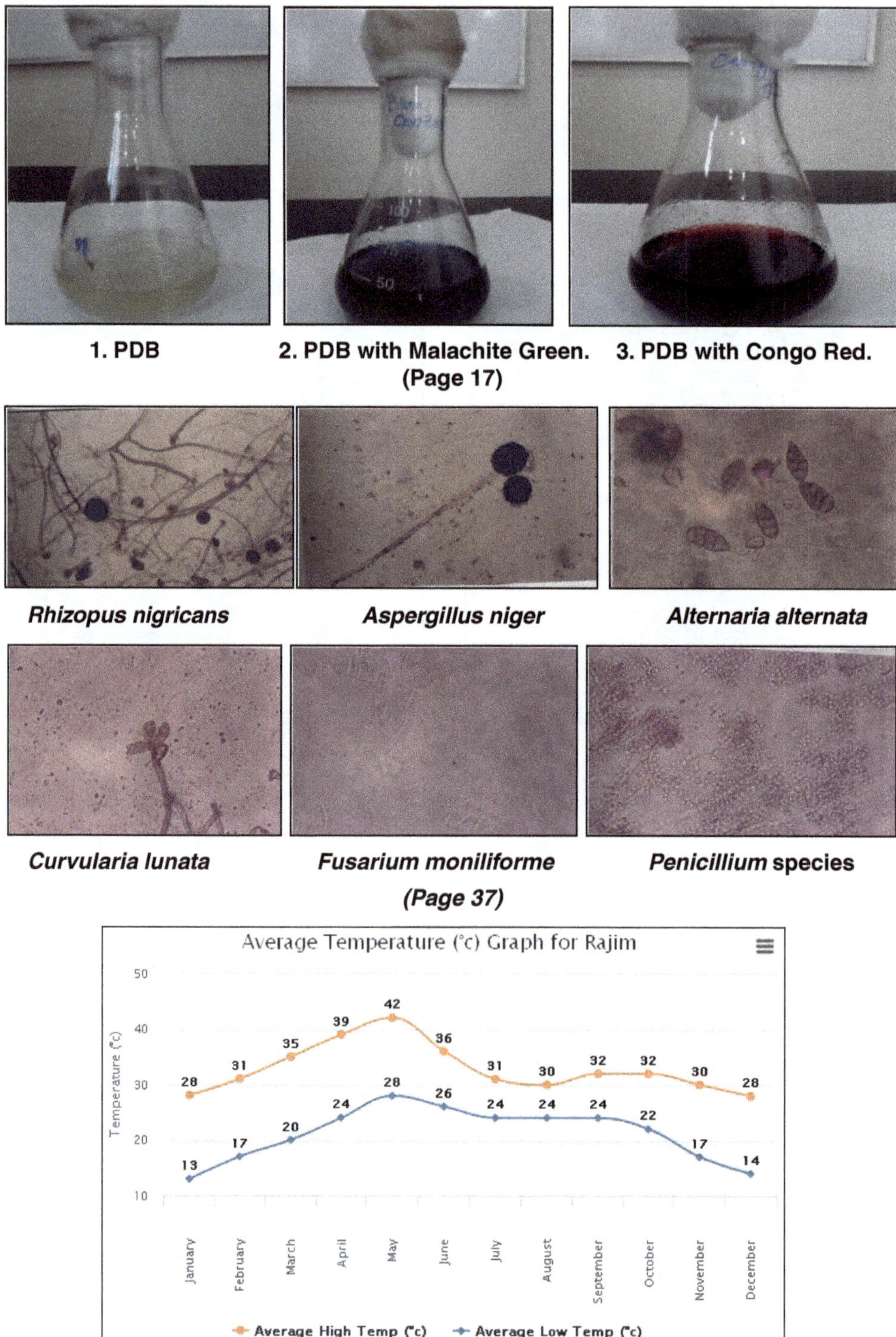

1. PDB **2. PDB with Malachite Green.** **3. PDB with Congo Red.**

(Page 17)

Rhizopus nigricans ***Aspergillus niger*** ***Alternaria alternata***

Curvularia lunata ***Fusarium moniliforme*** ***Penicillium* species**

(Page 37)

Figure 7.3: Average High/Low Temperature for Rajim, India. (Page 43)

Rungia pectinata **(L.) Ness**

Tephrosia purpurea **(Linn.) Pers.**

Spilanthus acmella **F. Linn.**

Asteracantha longifolia **(L.)**

Figure 7.5: Some Weeds. (Page 49)

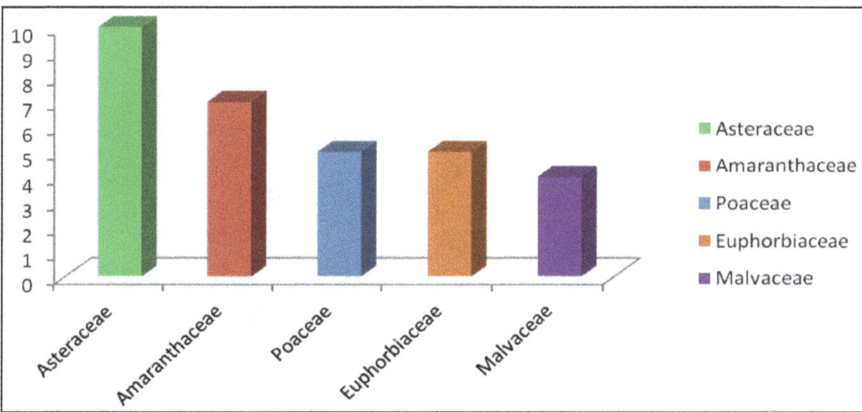

Figure 7.6: Five Dominant Families. (Page 50)

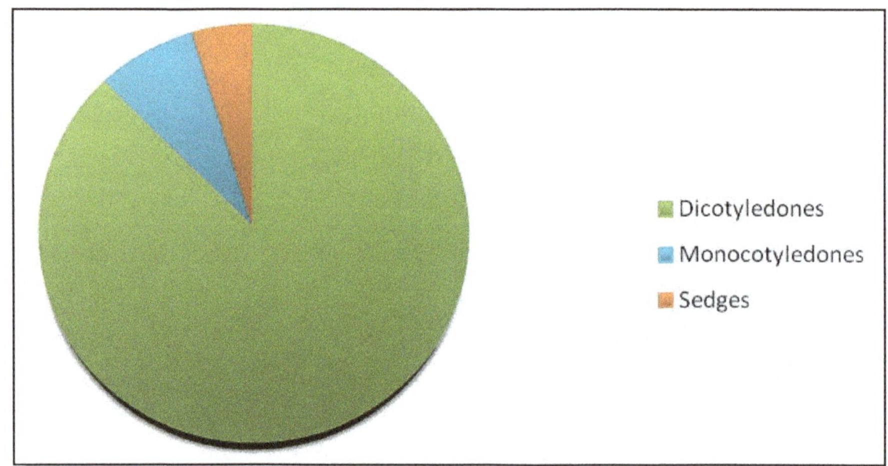

Figure 7.7: Distribution of Weed Species. (Page 50)

***Azadirachta indica* (Page 56)**

***Butea monosperma* (Page 56)**

***Carica papaya* (Page 57)**

***Caryota urense* (Page 57)**

Cassia fistula (Page 57)

Cycas revoluta (Page 57)

Emblica officnalis (Page 57)

Euphorbia lactea (Page 57)

Algal Culture by Open Pond. (Page 74)

Photo-bioreactors. Page 75

Photo-bioreactors. Page 80

Figure 13.1: *Mimosa hamata* Plant. (Page 88)

Figure 13.2: Powdered *Mimosa hamata* Leaves. (Page 88)

www.ingramcontent.com/pod-product-compliance
Ingram Content Group UK Ltd.
Pitfield, Milton Keynes, MK11 3LW, UK
UKHW021011290726
14059UKWH00001BA/72